SCIENTIFIC AMERICAN EXPLORES BIG IDEAS

Extraterrestrial Physics

Published in 2024 by Scientific American Educational Publishing
in association with **The Rosen Publishing Group**
2544 Clinton Street, Buffalo NY 14224

Contains material from Scientific American®, a division of Springer Nature America, Inc., reprinted by permission, as well as original material from The Rosen Publishing Group®.

First Edition

Scientific American
Lisa Pallatroni: Project Editor

Rosen Publishing
Daniel R. Faust: Compiling Editor
Michael Moy: Senior Graphic Designer

Cataloging-in-Publication Data
Names: Scientific American, Inc.
Title: Extraterrestrial physics / edited by the Scientific American Editors.
Description: First Edition. | New York : Scientific American Educational Publishing, 2024. | Series: Scientific American explores big ideas | Includes bibliography and index.
Identifiers: ISBN 9781725350182 (pbk.) | ISBN 9781725350199 (library bound) | ISBN 9781725350205 (ebook)
Subjects: LCSH: Astrophysics–Juvenile literature. | Cosmology–Juvenile literature. | Astronomy–Juvenile literature. | Universe–Juvenile literature.
Classification: LCC QB461.E987 2024 | DDC 523.01–dc23

Manufactured in the United States of America
Websites listed were live at the time of publication.

Cover: Elena11/Shutterstock.com

CPSIA Compliance Information: Batch # CWSA24.

For Further Information contact Rosen Publishing at 1-800-237-9932.

CONTENTS

INTRODUCTION

Astronomy is the oldest of the natural sciences. Since prehistory, humans have looked up and studied the sun, moon, and stars. Early cultures believed these celestial objects were associated with gods, spirits, and other supernatural beings. They correctly related the movement of these objects to natural phenomena like rain and the seasons. The origins of modern astronomy can be traced back to the ancient kingdoms of Mesopotamia. For thousands of years, astronomers could only study objects that they could see with their naked eyes. New inventions and technological advancements changed all of that. Telescopes and computers made it possible to study objects in deep space. Modern astronomers can do a lot more than simply observe distant stars, pulsars, and supernovae. They can measure the levels of radiation and the changes in gravity that could indicate the presence of unobservable phenomena like black holes or subatomic particles like bosons.

The articles selected for this volume explore the celestial phenomena, both proven and theoretical, being studied by today's physicists. You've probably heard of the Big Bang and its role in the creation of the universe. Section 1, "Big Beginnings," explores the origins of our universe and how physicists are studying distant galaxies to better understand how the universe began. In Section 2, "Black Holes," you'll get up close and personal with the objects that nothing can escape, not even light. Section 3, "Dark Matter," examines the mysterious theoretical substances known as dark matter and dark energy. String theory and quantum entanglement are the focus of Section 4, "Entanglement." Section 5, "Gravitational Waves," looks at the way physicists use the behavior of gravity to search for unobservable phenomena. Section 6, "Ongoing Mysteries and Unanswered Questions," looks to the future of astronomy and physics, and explores some of the potential discoveries just over the horizon.

Section 1: Big Beginnings

The Universe Began with a Bang, Not a Bounce, New Studies Find

By James Riordon

How did the universe start? Did we begin with a big bang, or was there a bounce? Might the cosmos evolve in a cycle of expansion and collapse, over and over for all eternity? Now, in two papers, researchers have poked holes in different models of a so-called bouncing universe, suggesting the universe we see around us is probably a one-and-done proposition.

Bouncing universe proponents argue that our cosmos didn't emerge on its own out of nothing. Instead, advocates claim, a prior universe shrunk in on itself and then regrew into the one we live in. This may have happened once or, according to some theories, an infinite number of times.

So which scenario is correct? The most widely accepted explanation for the history of the universe has it beginning with a big bang, followed by a period of rapid expansion known as cosmic inflation. According to that model, the glow left over from when the universe was hot and young, called the cosmic microwave background (CMB), should look pretty much the same no matter which direction you face. But data from the Planck space observatory, which mapped the CMB from 2009 to 2013, showed unexpected variations in the microwave radiation. They could be meaningless statistical fluctuations in the temperature of the universe, or they might be signs of something interesting going on.

One possibility is that the CMB anomalies imply that the universe didn't emerge out of nothing. Instead it came about after a prior universe collapsed and bounced back to create the space and time we live in today.

Bouncing universe models can explain these CMB patterns as well as account for lingering quibbles about the standard description of the universe's origin and evolution. In particular, the big bang model of the universe begins with a singularity—a point that appeared out of nothing and contained the precursors of everything in the universe in a region so small that it had essentially no size at all. The idea is that the universe grew from the singularity and, after inflation, settled into the more gradually expanding universe we see today. But singularities are problematic because physics, and math itself, doesn't make sense when everything is packed into a point that's infinitely small. Many physicists prefer to avoid singularities.

One bouncing model that averts singularities and makes the CMB anomalies a little less anomalous is known as loop quantum cosmology (LQC). It relies on a bridge between classical physics and quantum mechanics known as loop quantum gravity, which posits that the force of gravity peters out at very small distances rather than increasing to infinity. "Cosmological models inspired by loop quantum gravity can solve some problems," says University of Geneva cosmologist Ruth Durrer, "especially the singularity problem." Durrer co-authored one of the two new studies on bouncing universes. In it, she and her colleagues looked for astronomical signs of such models.

In an LQC model, a precursor to our universe might have contracted under the force of gravity until it became extremely compact. Eventually quantum mechanics would have taken over. Instead of collapsing to a singularity, the universe would have started to expand again and may even have gone through an inflationary phase, as many cosmologists believe ours did.

If that happened, says physicist Ivan Agullo of Louisiana State University, it should have left a mark on the universe. Agullo, who was not affiliated with either of the recent analyses, has proposed that the mark would turn up in a feature in the CMB data known as the "bispectrum," a measure of how different portions of the universe would have interacted in a bouncing scenario. The bispectrum would not be apparent in an image of the CMB, but it would show up in analyses of the frequencies in the ancient CMB microwaves.

"If observed," Agullo says, the bispectrum "would serve as a smoking gun for the existence of a bounce instead of a bang."

Agullo's group previously calculated the bispectrum as it would have appeared shortly after a cosmic bounce. Durrer and her colleagues took the calculation further, but when they compared it with the present-day Planck CMB data, there was no significant sign of a bispectrum imprint.

Although lots of other bouncing cosmos models may still be viable, the failure to find a significant bispectrum means that models that rely on LQC to deal with the anomalies in the CMB can be ruled out. It's a sad result for Agullo, who had high hopes of finding concrete evidence of a bouncing universe. He still considers many bouncing universe models viable, however. And Paola Delgado, a cosmology Ph.D. candidate at Jagiellonian University in Poland, who worked on the new analysis that was co-authored by Durrer, says there's one potential upside. "I heard for a long time that [attempts to merge quantum physics and cosmology] cannot be tested," Delgado says. "I think it was really nice to see that for some classes of models, you still have some contact with observations."*

Ruling out signs of an LQC-driven cosmic bounce in Planck data means the CMB anomalies remain unexplained. But an even larger cosmic issue lingers: Did the universe have a beginning at all? As far as advocates of the big bang are concerned, it did. But that leaves us with the inscrutable singularity that started everything off.

Alternatively, according to theories of so-called cyclic cosmologies, the universe is immortal and is going through endless bounces. Although a bouncing universe may experience one or more cycles, a truly cyclic universe has no beginning and no end. It consists of a series of bounces that go back for an infinite number of cycles and will continue for an infinite number more. And because such a universe doesn't have a beginning, there's no big bang and no singularity.

The study that Durrer and Delgado co-authored doesn't rule out immortal cyclic cosmologies. Plenty of theories describe such a bouncing universe in ways that would be difficult or impossible to distinguish from the "big bang plus inflation" model by looking at Planck CMB data.

But a critical flaw lurks in the idea of an eternally cycling universe, according to physicist William Kinney of the University at Buffalo, who co-authored the second recent analysis. That flaw is entropy, which builds up as a universe bounces. Often thought of as the amount of disorder in a system, entropy is related to the system's amount of useful energy: the higher the entropy, the less energy available. If the universe increases in entropy and disorder with each bounce, the amount of usable energy available decreases each time. In that case, the cosmos would have had larger amounts of useful energy in earlier epochs. If you extrapolate back far enough, that implies a big bang–like beginning with an infinitely small amount of entropy, even for a universe that subsequently goes through cyclic bounces. (If you're wondering how this scenario doesn't violate the law of conservation of energy, we're talking about available energy. Although the total amount of energy in the cosmos remains static, the amount that can do useful work decreases with increasing entropy.)

New cyclic models get around the problem, Kinney says, by requiring that the universe expands by a lot with each cycle. The expansion allows the universe to smooth out, dissipating the entropy before collapsing again. Although this explanation solves the entropy problem, Kinney and his University at Buffalo co-author Nina Stein calculated in their recent paper that the solution itself ensures that the universe is not immortal. "I feel like we've demonstrated something fundamental about the universe," Kinney says, "which is that it probably had a beginning." That implies a big bang occurred at some point, even if that event happened many bouncing universes ago, which in turn suggests that it took a singularity to get everything going in the first place.

Kinney's paper is the latest in the debate over cyclic universes, but proponents of a universe without beginning or end have yet to respond in the scientific literature. Two leading proponents of a cyclic universe, astrophysicists Paul Steinhardt of Princeton University and Anna Ijjas of New York University, declined to comment for this article. If the history of the debate is any indication, though, we may soon hear of a work-around to counter Kinney's analysis.

Cosmologist Nelson Pinto-Neto of the Brazilian Center for Physics Research, who has studied bouncing and other cyclic models, agrees that the Planck data likely rule out a bounce under loop quantum cosmology, but he's more sanguine on the question of a cyclic universe. "Existence is a fact. We are all here and now. Nonexistence is an abstraction of the human mind," Nelson says. "This is the reason I think that a [cyclic universe], which has always existed, is simpler than one that has been created. However, as a scientist, I must be open to both possibilities."

**Editor's Note (6/1/23): This paragraph was edited after posting to better clarify Ivan Agullo's position on bouncing universe models.*

About the Author

James Riordon is a freelance writer based in Clarksville, MD. His book about the neutrino, Ghost Particle *(with co-author Alan Chodos), was published in 2023.*

Colliding Dwarf Galaxies Reveal a Glimpse of the Early Universe

By Meghan Bartels

For the first time, astronomers think they've spotted two pairs of merging dwarf galaxies, with each pair sporting a duo of soon-to-collide black holes. The observations could reveal new details about the early formation of large galaxies like our own Milky Way, as well as the supermassive black holes that lurk at their centers.

Despite their relatively small size, dwarf galaxies pack a hefty scientific punch. Thought to be the most abundant galaxy type during the first several hundred million years after the big bang, when the expanding universe was smaller and denser than it is now, dwarf galaxies must have been progenitors of the bigger objects we see today. Across eons, large galaxies gradually assembled via countless collisions of their dwarfish kin, scientists presume. But whether these cosmic building blocks also contained sizable central black holes has been something of an open question. Now, however, astronomers believe they've discovered a clue. They have found evidence of bulky black holes feeding on gas and dust in what seem to be two colliding pairs of dwarf galaxies. The team described the two candidate systems in a paper published last month in the *Astrophysical Journal*.

Researchers have previously found examples of merging dwarf galaxies, some of which even showed signs of harboring a black hole in one of the galaxies, says lead study author Marko Mićić, a doctoral student in astronomy at the University of Alabama. But never before has anyone seen dual black holes in colliding dwarfs—the key to why the new finds are so intriguing. Each of the two systems Mićić and his colleagues have studied, he says, is "a complete analogue of events that probably happened in the very early universe."

And perhaps best of all, one of the two pairs is relatively close in cosmic terms—a mere 760 million light-years away, making it a good target for further in-depth observations. That isn't to say that those observations will be easy: because dwarf galaxies contain only a vanishing fraction of the stars and mass held in a typical Milky Way–sized galaxy, they're also fainter and more difficult to study than their larger counterparts. And they don't necessarily behave in the same way larger galaxies do.

"Dwarf galaxies usually look like little train wrecks–they're messy and chaotic," says Sabrina Stierwalt, a dwarf galaxy specialist at Occidental College, who was not involved in the new research. "They're very cool laboratories."

To spot the two intriguing mergers, Mićić and his colleagues used archival data from a trio of instruments: NASA's Chandra X-ray Observatory, the space agency's Wide-field Infrared Survey Explorer (WISE) space telescope, now called NEOWISE, and the Canada-France-Hawaii Telescope (CFHT), which is located in Hawaii.

Although black holes themselves are invisible, they can spark the universe's brightest fireworks when they devour surrounding material. Infalling matter swirls around them at nearly the speed of light, heating up to emit powerful glows that shine across the cosmos in multiple wavelengths of light. Chandra saw that glow in the x-ray part of the electromagnetic spectrum, while WISE glimpsed it in infrared. Seeing twin glows for each merging dwarf pair in both Chandra and WISE data suggests–but doesn't confirm–the presence in each of two feeding, "active" black holes, the researchers say. CFHT provided images of the dwarf galaxies themselves, revealing starry evidence of the likely intergalactic smashups, as well as optically bright blobs aligned with the x-ray and infrared glows.

The closer candidate merger appears to be in the late stages of coalescence, with one dwarf galaxy accompanied by a flamboyant, clumpy tail of stellar debris that is probably the shredded remnants of the other. That tail inspired the scientists to dub the system Mirabilis, named after *Loddigesia mirabilis*, a hummingbird species that boasts two long, thin tail plumes that end in a Ping-Pong paddle of feathers.

The second candidate is much farther away—about 3.2 billion light-years—and is earlier in the merging process. Because the two dwarf galaxies are more widely separated in space and still largely intact, the scientists bestowed both with names, calling one Elstir and the other Vinteuil in a nod to characters in Marcel Proust's novel *In Search of Lost Time*. This merger, Mićić says, will likely take hundreds of millions of years to complete.

Mićić and his colleagues are obtaining additional observations of the two pairs, particularly in radio and ultraviolet light. Some of these follow-up observations could eliminate small-but-significant lingering uncertainties about whether the galaxies are really dwarfs rather than the twisted wreckage of giant galaxies wrenched apart by gravitational forces in the galaxy clusters they inhabit. Further scrutiny could also shore up the case for the bright glows at their cores really coming from voracious black holes and even help them calculate the masses of these hidden giants.

If the discovery pans out, Stierwalt says, it could become a crucial component for solving the greater cosmic puzzle of how exactly galaxies and black holes have emerged and coevolved across the universe's history. "This would be an important piece of evidence that some of the processes that happen in massive galaxy mergers also happen in dwarf galaxy mergers," she says, "like the fact that you can have a merging of black holes."

About the Author

Meghan Bartels is a science journalist and news reporter for Scientific American *who is based in New York City.*

Ancient Galaxy Clusters Offer Clues about the Early Universe

By Arianna S. Long

Just like trees, people and stars, galaxies have life cycles. A galaxy is born when enough gas and stars coalesce to form a coherent structure–perhaps it starts as one cloud of gas and slowly gathers mass, or maybe it builds up from the collision of two or more clouds. Either way, once formed, a galaxy spends its lifetime making stars, using its reservoirs of gas to create tiny furnaces where nuclear fusion burns elements to release light and energy. A galaxy deemed "alive" shines strongly in ultraviolet light, a signal of young, bright and hot stars. As those stars age, their light changes from hot and blue to cool and yellow or red. When a galaxy contains mostly yellow and red stars and emits little to no ultraviolet light, we consider it retired, or "red-and-dead." Eventually, if massive enough, it becomes a spheroidal blob, known as an elliptical galaxy, that will likely never birth a new star again.

All around us in nearby space–within, say, 300 million to 600 million light-years–astronomers see dead or dying elliptical galaxies gathered in great ensembles called galaxy clusters. These clusters hold the fossilized remains of the most massive galaxies ever formed–hundreds to thousands of them slowly dancing around one another, gravitationally bound forever in their permanent graves.

But galaxy clusters present a problem for astronomers. Most clusters seem to have been established by the time the universe was only half of its current age. That means the galaxies within those clusters must have birthed most of the stars they contain early in cosmic history. It appears that these galaxies grew to the size of the Milky Way or larger but up to 10 billion years more quickly. The young galaxy clusters, called protoclusters, where these galaxies formed must have been incredibly violent and active places, full of galaxies producing stars at a furious pace. Our current understanding

of physics cannot quite explain how they could have grown so big so quickly.

Only recently have astronomers had the telescopic tools necessary to find protoclusters, which are very distant (their light often travels 10 billion years or more to reach us) and frequently hide their most massive galaxy members behind dust. In the past few years scientists have discovered two protoclusters that are providing an unprecedented window into cluster growth. Follow-up observations have revealed that they are, in fact, active and huge—so huge that they challenge our understanding of galaxy formation. If we can solve the riddle posed by galaxy clusters, we may redefine our understanding of the evolution of the universe.

Hunting for Starbursts

The most common type of star-forming galaxy produces roughly one to tens of suns' worth of stars a year. These are often called normal star-forming galaxies. The Milky Way is in this class. Normal star-forming galaxies are metaphorical tortoises, forming stars slowly and steadily over the course of 10 billion years or so, remaining blue and disklike and depleting their reservoirs of fresh gas (fuel for new stars) at a leisurely pace.

Galaxies that produce hundreds to thousands of stars every year are known as starburst galaxies. These are the hares in galaxy evolution. In perhaps 300 million years at most, these galaxies burst into existence, form as many stars as possible as quickly as possible and, in a cosmic blink of an eye, run out of fuel. Starbursts live fast and die young. Astronomers think that they are the best candidate ancestors for the massive, dead elliptical galaxies we see in clusters today.

It stands to reason that if we looked deep enough into space, we would find protoclusters filled with starburst galaxies—tomorrow's clusters of dead galaxies. Yet finding starburst galaxies in protoclusters has proved challenging. Until recently, most of our methods for spotting clusters were developed to preferentially

select dying elliptical galaxies or the hot gas that pervades the space between them. Elliptical galaxies and hot intracluster gas appear at the later stages of galaxy cluster evolution, so we need new methods to find their bluer, more star-forming infant counterparts. To make things more difficult, protoclusters are often spread far apart on the sky because the galaxies have yet to fully coalesce into the dense structures we see today. When our most famous and precise telescopes have cameras that span only the width of a pencil (the Hubble Space Telescope, for instance), it is not surprising that we cannot piece together protocluster puzzle pieces that are spread across the sky at distances more than 100 times greater than our telescope's field of view.

Other methods of searching, such as systematically surveying large swaths of the sky, tend to miss starburst galaxies because they are often obscured by dust. The exceptional stellar growth in starbursts generates an overabundance of heavy metals that are produced in the explosive deaths of stars. Once dispersed into space, heavy elements such as iron, carbon and gold collide to form complex dust molecules that absorb and obscure ultraviolet and optical light. Think of the reddening sun during wildfire season: dust dims hotter, more energetic blue light while letting redder light sneak through. The result is that starburst galaxies are nearly invisible when viewed with optical and ultraviolet telescopes, but they shine like beacons when viewed in the cooler infrared spectrum.

All of this means that until recently, the tools to find and study protoclusters usually missed a key population of galaxies. From the late 1990s through the early 2010s the Submillimeter Common-User Bolometer Array, the Herschel Space Observatory, the South Pole Telescope and the Spitzer Space Telescope revolutionized our understanding of the dust-obscured universe by unveiling millions of galaxies that were previously invisible. Starting about 15 years ago, astronomers began studying the clustering properties of dusty starbursts, and they found that these powerhouses live preferentially near other large and actively star-forming galaxies. But the state of technology was still behind our ambitions; the resolution of infrared

and millimeter telescopes was still so low that multiple galaxies would get blended into one large object, even if those galaxies were far apart but lay along the same line of sight. The age of the infrared universe was here, but we needed sharper and more sensitive instruments to fully comprehend what we were seeing.

Finally, in 2013, the Atacama Large Millimeter/submillimeter Array (ALMA) arrived. High in the Chilean desert, this collection of nearly 70 radio dishes works together as a single telescope, reaching resolutions up to 600 times sharper than that of the Herschel telescope. ALMA has transformed many corners of astronomy, including galaxy evolution. (I know several people with tattoos dedicated to this telescope.) The observatory is excellent at detecting dusty, gaseous stellar nurseries throughout star-forming galaxies. With it, astronomers have discovered systems that are both shocking and exciting.

Surprising Behemoths

In 2018 two separate teams of astronomers used ALMA to study the brightest infrared objects they could find in the distant universe. Each team discovered a different conglomeration of dusty, starbursting galaxies that were previously blended together, hiding as one in surveys taken by the first generation of infrared telescopes. SPT2349-56, a group of 14 galaxies, and the Distant Red Core (DRC), a group of 10 galaxies, were both found growing and thriving in different corners of the universe, when the cosmos was only 10 percent of its current age. We see both these budding protoclusters undergoing extreme bursts of star formation—each group birthing nearly 10,000 times as much stellar mass a year as the Milky Way does—across volumes only half the size of our Local Group (which includes our own galaxy plus Andromeda and several smaller galaxies). Estimates of the protoclusters' individual gas reservoirs tell us that if these galaxies continued to form stars at such excessive rates, they would exhaust their fuel supply in just a few hundred million years and become the massive red-and-

dead elliptical galaxies that are ubiquitous in fully grown clusters. Moreover, they would complete this cycle well before the present era.

The discovery of these two dust-obscured protocluster cores presented a promising new lens for studying cluster growth, but we were still missing an important part of the picture. The best way to "weigh" a galaxy is to measure the light from its adult star population, which requires data from across the electromagnetic spectrum. But until recently, all observations of protoclusters living in the first two billion years of the universe had been conducted within a narrow energy spectrum (in either the optical or the infrared). Then, in September 2018, my colleagues and I were able to observe, for the first time, ultraviolet and optical emission from a dusty, starbursting protocluster as seen 12 billion years ago: the Distant Red Core. Using the Hubble Space Telescope, the Gemini Observatory and the Spitzer Space Telescope, we captured the multiwavelength perspective necessary to more deeply understand this structure's past and future.

Waiting for Hubble Space Telescope data can be anticlimactic. You know the day that your patch of the sky is scheduled for observation, but you have no idea when you will actually receive your data: you just have to wait for an e-mail notification to tell you to check the archive. The day our protocluster observations were scheduled, I checked my e-mail what felt like every two minutes. I was disappointed when it was time to go to bed and nothing had arrived in my in-box.

The following morning, against my partner's protests, I rolled out of bed immediately and went straight to my computer to see if the data had at last come in. Fortunately, it had been delivered a few hours past midnight. I commenced the download, dancing impatiently like a child waiting for her turn to unwrap presents. Finally, I opened the image. No words can describe how it feels to be the first person to glimpse a part of the universe that no one else has seen. I felt compelled to take a moment to inspect each star and galaxy in the field, to acknowledge their existence. Eventually I snapped back to myself and zoomed in on the part of the sky I was interested in. I saw something remarkable.

This little region of space is violent. At least half of the galaxies there were so messily shaped they must have recently crashed into other nearby galaxies or were still in the process of doing so. When we measured the population of adult stars in these galaxies, we found something incredible–so incredible that it may pose a problem for our current understanding of the universe. Already at this early era, some of the Distant Red Core galaxies had formed three times more stars than our own Milky Way has–but in just a fraction of the time. Yet simulations of the universe based on known physics struggle to produce galaxies this massive so early on. This incongruity between simulation and observation exacerbates a problem that we have known about since the discovery of dusty, star-forming galaxies. Modeling the extreme pace and density of star formation seen in starbursts is difficult because physics predicts the simulated galaxies should either shred themselves apart or heat up so much that they blow out all of the fresh gas needed to grow large enough to match what we see today.

The protocluster as a whole presents another problem: it is shockingly massive. When I first measured it, I could not believe my numbers. I knocked on doors in my department to make sure I was doing the calculations right. Two weeks later I brought the results to a conference to show my collaborators. One said, "There must be a bug in your code." Another asked, "Are you sure you're not double counting somewhere?" (Turns out there was a small bug in my code, but it was not enough to explain the huge measurement.) Eventually, after double-checking my calculations and trying out different methods, the measurements became undeniable. The Distant Red Core seems to be too big for our universe. We do not know how it could have gotten so large in such a short time.

To better understand its bulk and how much mass was in the form of stars, we zeroed in on the size of the dark matter halo around this protocluster. Dark matter is the most abundant form of matter in any given galaxy and in the universe as a whole. All galaxies and clusters are thought to be surrounded by blobs, or halos, of this mysterious stuff. And although it is invisible and poorly understood, dark matter leaves a clear gravitational signal. There are a variety of ways to infer the amount of dark matter in a given astronomical object, and to cover those methods would require an additional article (or five).

Suffice it to say, we weighed the dark matter component of the Distant Red Core, and according to our simulations, it contains nearly the largest allowable halo mass at that period in the history of the universe. This apparent overabundance of dark matter means that the DRC may be so large that it violates the laws of our universe as we understand them. When we fast-forward our simulations to estimate what the DRC may look like after evolving 12 billion years to the present day, we find that it may grow to be larger than the largest known galaxy cluster, El Gordo. Although we have a healthy margin of error on our dark matter calculation (meaning it could be overestimated), the discrepancy looks even worse when we consider the fact that our observations are capturing only a small percentage of the likely galaxy cluster members; there are probably more galaxies in the DRC that were simply out of the narrow field of view of our telescopes and thus not included in our calculations. This mismatch most likely will grow as we continue surveying and studying this protocluster.

Rethinking the Time Line

Our investigation of the Distant Red Core, along with the discoveries of other potentially similar protoclusters, forces us to reconsider our understanding of galaxy cluster formation. Because the galaxies in clusters are likely to be some of the first galaxies ever, we must determine how such massive objects can form so quickly. Doing so is not just an issue of constraining the physical mechanics and chemistry of star formation inside the first galaxies. It is also a matter of investigating the timing of the conditions that lead dark matter to gravitationally collapse into halos, seeding galaxies. Is it possible that galaxies and structure began forming earlier in the universe than we thought? What does that mean for our understanding of the formation of the first elements? Could these galaxies have forged the right ingredients to build stars with habitable planets around them—and perhaps hosted some of the first forms of life in the universe?

Some of these questions probably will not be answered during my lifetime, but I and other astrophysicists are working hard and fast to address the others. Already we are carrying out more observations of these known protoclusters across the electromagnetic spectrum. We are also developing new methods for identifying large samples of dusty protocluster candidates. With more examples, we may be able to determine whether protoclusters such as the Distant Red Core are examples of a common, yet previously invisible, phase of galaxy evolution that all clusters go through or just rarities. Observers and theorists are forming new collaborations to learn how early in the history of the universe conditions were right for protoclusters akin to those we have discovered—pockets of space overdense with tremendous rates of star formation and outsize masses.

The best way to test our physical models is to look at extremes. In the next few years these colossal congregations of exceptional galaxies will be putting humanity's grasp on the cosmos to the test.

About the Author

Arianna S. Long is a graduate student in physics at the University of California, Irvine, who investigates the history of galaxy clusters.

Giant Galaxies from the Universe's Childhood Challenge Cosmic Origin Stories

By Robin George Andrews

Recently an international team of astronomers traveled back in time to when our universe was just 1.8 billion years old. They did not go directly, of course, but settled for the next best thing: gathering 17 hours' worth of starlight from a single small patch of the distant cosmos with the Large Binocular Telescope Observatory atop Mount Graham in southeastern Arizona. Such clock-rewinding virtual voyages are routine in astronomy—light's finite speed ensures that the deeper into space you see, the further back in time you gaze. And many observatories around the globe can gather faint photons from ancient skies. But this particular cosmic jaunt concerned something special—even disturbing: an abnormally hefty elliptical galaxy dubbed C1-23152. This egg-shaped aggregation of stars is so outsize that it defies conventional models of its origins. Simply put, C1-23152 seems to be too big to fit the early universe.

It is thought that the first galaxies were relatively minuscule, clumping together from smaller building blocks bit by bit and only reaching gargantuan sizes after billions of years of growth. Boasting an estimated 200 billion solar-mass stars, C1-23152 tips and then overturns the scales for this scenario. And it is not alone. Over the past decade, astronomers have discovered several very ancient, very big galactic behemoths. In 2017, for example, a pair of extremely large galaxies—one capable of churning out 2,900 solar masses of stars per year—were found to exist less than 800 million years after the big bang. In 2019 a family of 39 huge galaxies—each a star factory manufacturing perhaps 200 solar-mass stars per year—were found zipping through the universe within two billion years of its birth.

Does this ever increasing number of venerable vast objects threaten to bring down the traditional model of galaxy formation?

"The trick here is: How many do you have?" says Marcel Neeleman, an astronomer at the Max Planck Institute for Astronomy in Heidelberg, Germany, who was not involved with the new study. A handful will not matter; the universe is big enough that odd things will crop up every now and then. But if future, increasingly advanced telescopes manage to find far more of them, then perhaps these colossal galaxies from the universe's childhood may break our understanding of the cosmos.

A Long Time Ago in a Galaxy Far, Far Away

What has become the widely accepted model of galaxy formation is largely gleaned from simulations of cosmic evolution that reproduce our observations of the local universe–the stuff we can see near the Milky Way.

After the big bang, the cosmos expanded and stretched out fairly evenly in all directions. But, Neeleman says, you get "tiny density variations in the fabric of the universe." These variations are home to clumps of dark matter, a substance that emits little, if any, electromagnetic radiation. As such, dark matter has yet to be directly detected, but observations of galaxies indicate that this invisible mass produces its own gravitational pull. That means that these dark matter clumps attract "ordinary" matter (the stuff we humans can detect and interact with), most of which is gas. The gas tumbles into these gravity wells and squashes together to trigger star formation. More matter continues to tumble into these ever expanding wells–called dark matter "halos" by astronomers–gradually forming bigger and bigger structures over the 13.8-billion-year lifetime of the universe. This process should more or less create the distribution of galaxies we see today, says Paolo Saracco, an astronomer at Italy's National Institute for Astrophysics and the lead author of a study reporting the recent observations of C1-23152.

That is why ancient massive galaxies are problematic. "For our current understanding of galaxy formation, we sort of built on the galaxies we knew at the time," says Coral Wheeler, an astronomer at California State Polytechnic University, Pomona, who was not involved with the new study. These galaxies did not include the very old, small or big ones. Looking further back in time with increasingly powerful telescopes began to reveal these apparent outliers. And as the tally of anomalous entities rose, astronomers started wondering if their models needed to expand to make room for them or if those models would buckle and break under the strain.

As reported in the *Astrophysical Journal* in December 2020, Saracco's team managed to extract some juicy details out of C1-23152. Light from far-distant cosmic regions is stretched by the expanding universe as it travels to Earth. The more it is stretched, the greater its shift towards the longer-wavelength "redder" section of the electromagnetic spectrum. This "redshift" of C1-23152's starlight indicates that it appeared 12 billion years ago, way back in the universe's youth. The fact that this galaxy is both ancient and massive alone is problematic enough for traditional slowly-but-surely models of galaxy formation. But it did not just appear fully formed. Saracco and his team's real breakthrough was to trace C1-23152's history of star formation from across the universe.

The key to that breakthrough was seeing the giant galaxy's spectrum–a rainbowlike measurement of the various wavelengths, or colors, that an object emits or absorbs. Particular color combinations distinguish specific elements, which means this spectral symphony can be used to determine the composition of a galaxy's stars. Using that power, Saracco says, "for the first time, we derived, with very good accuracy, the mean age of the stellar population inside [C1-23152] and the time necessary to form those stars."

The number of elements in C1-23152 that were found to be heavier than hydrogen and helium–which astronomers collectively refer to as "metals"–hinted at its strangeness. Metals are produced by star formation, which jettisons them into a galaxy's interstellar medium through supernovae–making them available for next-generation stars to use. More metals equal more cycles of star formation, and it took present-day massive galaxies many billions of years to become metal-rich. C1-23152's spectrum revealed the galaxy to be a veritable metal bonanza back in its early days, which means it made a *lot* of stars very rapidly not long after it first formed.

How rapidly? The spectral features of stars can answer that question, too, because they reveal which ones have elements typical of younger or older stars. The youngest stars in C1-23152 are roughly 150 million years old. The most ancient are about 600 million years old. That means the galaxy made some 200 billion solar masses in just a half-billion years—a rate of 450 stars per year, more than one per day. The figure is almost 300 times greater than estimates of the Milky Way's current output. If most galaxies are slow-burning log fires, with new flames popping up every so often, C1-23152 is a gasoline-soaked bonfire.

C1-23152 and its similar cousins present astronomers with a potentially model-breaking conundrum: How can massive galaxies be assembled and set alight so quickly so early on? For now the answer, in short, is that they can't.

Growing the Universe in a Box

For some time, simulations have struggled to grow these ginormous galaxies. But that does not mean they simply cannot do so. Instead the trouble may lie in how they are programmed.

"When you run a simulation, there's a trade-off between how big of a volume you want to simulate and how much detail you can simulate because of the computer power you have or don't have," says Ben Forrest, an astronomer at the University of California, Riverside, and a co-author of the new study. If these ancient massive galaxies are rare, perhaps we are not using big enough boxes to give one the chance to pop up. "Maybe some of the simulations aren't really covering enough volume," he says.

Quickly tweaking them to spawn mega galaxies from the early eras of cosmic time is not easy either. "It takes a long time to rerun them. If you want to change something, you've got to be pretty sure that's right and that's what you want to do," Forrest says.

Some of the latest iterations of simulations, with better data and computing power, do predict these massive galaxies to exist in small numbers at early times, he adds. But unlike what is being observed in reality, they tend to still be making stars. Ancient galaxies, including C1-23152, abruptly shut off star formation after a productive peak—either because they run out of hydrogen and helium fuel or because the radiation shooting out from fresh crops of stars and other overzealous astrophysical sources cooks that gas and blasts it out of reach. Clearly, some ingredients are still missing from our virtual recipes, so we cannot rely on them for an explanation yet.

Scientists have found clues elsewhere that may account for these ancient mega galaxies. Anastasia Fialkov, a cosmologist at the University of Cambridge, who was not involved with the latest work, says that, unlike full-blown simulations, analytic physics calculations can "take into account the whole volume of the universe." And they suggest that a small number of dark matter halos capable of initiating star formation show up just 40 million years after the big bang.

That time is significantly earlier than the majority of dark matter halos that turn up later on in the youthful epochs of the universe—those thought to be responsible for seeding much of the galaxies we see today. Instead the halos that appeared 40 million years after the big bang would have been capable of seeding the beginnings of the ancient massive galaxies that would eventually become detectable via our telescopes. The early universe was also denser, Wheeler notes. That would make scooping up star-making hydrogen and helium around these primordial dark matter halos, and eventually galaxies, fairly effortless.

Another option, Neeleman says, is that a combination of things could have occurred. Rare hyperdense pockets of the universe would permit multigalaxy mergers very early on, while streams funneling gas into the hearts of galaxies could supercharge star formation.

In any event, the emergence of huge ancient galaxies is more easily explained if dark matter is cold. Here, "cold" means the dark matter moves relatively slowly. "Hot" dark matter would move at velocities approaching the speed of light. Generally speaking, the colder the dark matter, the easier it can condense into galaxy-seeding halos. This assumption may not necessarily be correct, but "cold dark matter is the simplest dark matter scenario that works," Fialkov says.

It is unclear which amalgam of these events, if any, best explains C1-23152's origins and evolution, let alone its colossal cousins. "This is not a special corner of the universe" we are looking at, Saracco says. But, importantly, nothing here threatens to overthrow the traditional slowly-but-surely model of galaxy formation, he says. These ancient, massive galaxies just represent another pathway for galaxies to take.

Back to the Future

The traditional model survives for now but only, in part, because few of these massive galaxies have been found. "We're dealing with small-number statistics," Forrest says. Scientists do not have a good grasp of the true amount of the behemoths, however. Until that changes, understanding what impact they have on our cosmic comprehension and how galaxies evolve in different ways will remain ambiguous.

Perhaps we have already seen many more of these old mega galaxies than we yet realize. For detailed studies, our telescopes are often drawn to the brightest massive but burnt out galaxies before their nature is revealed. Astronomers have spotted fainter objects with otherwise similar characteristics hanging about in the early universe, however, says Stijn Wuyts, an astronomer at the University of Bath in England, who was not involved with the recent work. They could turn out to be merely less massive galaxies or yet more ancient massive ones observed long after their star-forming prime. Are these objects dimmer candles closer to home or vast pyres farther afield?

As ever, more data are required. And several upcoming telescopes will aid us in this time-traveling galactic census.

First, suspicious bright splotches in the distant past need to be spotted. "If you want to get a bunch of candidates, then a wide field of view is great," Forrest says. The Nancy Grace Roman Space Telescope, formerly known as WFIRST and currently targeted for a 2025 launch, will have a field of view equivalent to 100 Hubble Space Telescopes: its wide, sensitive eyes will see plenty of possible ancient massive galaxies.

Those candidates will then need to be forensically examined by looking at their various spectra to determine their properties and confirm they are indeed such galaxies and not imposters. "Ideally, you want a really big telescope," Forrest says. "That gives you more collecting area–it's a bigger bucket for photons to go into from an object." Hawaii's Thirty Meter Telescope could be suitable if it is built, and the Extremely Large Telescope could fit the bill as well. The James Webb Space Telescope–which is finally launching this October after an abundance of delays–should work well, too. "It's not as big," Forrest says. "The bucket for the photons is a little bit smaller, but then you don't have to look through the atmosphere," so there is less interference to deal with.

Saracco is particularly excited for these upcoming next-generation magnifying glasses because they will do more than merely finding extremely distant objects. "We will be able to observe *inside* [a] galaxy, at single star-forming regions," he says. In other words, instead of a blurry picture of a galaxy's bulk characteristics, astronomers will get a more granular view–the difference between a rough sketch and a detailed painting–opening up a new chapter in our understanding of how galaxies form.

Until this help arrives, this scientific field will remain in its infancy. "There's so much uncertainty that goes into galaxy formation," Wheeler says.

It can be unnerving to chase monsters in the dark. They threaten the dogmas of the era, forcing us to expand our earlier models to fit them. And if those models stretch to the point of breaking, that's okay. "We want to challenge, in some way, the model," Wheeler says. "When things don't match, that's when it gets interesting."

About the Author

Robin George Andrews is a volcanologist, author and science writer based in London. His first book, Super Volcanoes: What They Reveal about Earth and the Worlds Beyond, *was published in November 2021.*

JWST's First Glimpses of Early Galaxies Could Break Cosmology

By Jonathan O'Callaghan

Rohan Naidu was at home with his girlfriend when he found the galaxy that nearly broke cosmology. As his algorithm dug through early images from the James Webb Space Telescope (JWST) late one night in July 2022, Naidu shot to attention. It had sifted out an object that Naidu recognized was inexplicably massive and dated back to just 300 million years after the big bang, making it older than any galaxy ever seen before. "I called my girlfriend over right away," Naidu says. "I told her, 'This might be the most distant starlight we've ever seen.'" After exchanging excited messages with one of his collaborators "with lots of exclamation marks," Naidu got to work. Days later they published a paper on the candidate galaxy, which they dubbed "GLASS-z13." The Internet exploded. "It reverberated around the world," Naidu says.

The discovery of this galaxy, just weeks into JWST's full operations, was beyond astronomers' wildest dreams. JWST–the largest, most powerful observatory ever launched from Earth–was built to revolutionize our understanding of the universe. Stationed 1.5 million kilometers away from earthly interference and chilled close to absolute zero by its tennis court–sized sunshade, the telescope carries a giant segmented mirror and exquisitely sensitive instruments that were designed to uncover details of cosmic dawn never before observed.

This is the scarcely probed era–no more than a few hundred million years after the big bang–in which the very first stars and galaxies coalesced. How exactly this process unfolded depends on exotic physics ranging from the uncertain influences of dark matter and dark energy to the poorly understood feedbacks among starlight, gas and dust. By glimpsing galaxies from cosmic dawn with JWST, cosmologists can test their knowledge of all these underlying phenomena–either confirming the validity of their best consensus models or revealing gaps in understanding that could herald profound new discoveries.

Such observations were supposed to take time; initial projections estimated the first galaxies would be so small and faint that JWST would find at best a few intriguingly remote candidates in its pilot investigations. Things didn't quite go as planned. Instead, as soon as the telescope's scientists released its very first images of the distant universe, astronomers such as Naidu (at the Massachusetts Institute of Technology) started finding numerous galaxies within them that, in apparent age, size and luminosity, surpassed all predictions. The competition for discovery was fierce: with each new day, it seemed, claims of yet another record-breaking "earliest-known galaxy" emerged from one research group or another. "Everyone was freaking out," says Charlotte Mason, an astrophysicist at the University of Copenhagen. "We really weren't expecting this."

In the weeks and months following JWST's findings of surprisingly mature "early" galaxies, theorists and observers scrambled to explain them. Could the bevy of anomalously big and bright early galaxies be illusory, perhaps because of flaws in analyses of the telescope's initial observations? If genuine, could they somehow be explained by standard cosmological models? Or, just maybe, were they the first hints that the universe is more strange and complex than even our boldest theories had supposed?

At stake is nothing less than our very understanding of how the orderly universe we know emerged from primordial chaos. JWST's early revelations could rewrite the opening chapters of cosmic history, which concern not only distant epochs and faraway galaxies but also our own existence here in the familiar Milky Way. "You build these machines not to confirm the paradigm but to break it," says JWST scientist Mark McCaughrean, a senior adviser for science and exploration at the European Space Agency. "You just don't know how it will break."

Deep Looks for Cosmic Dawn

One might say JWST's observations of early galaxies have been billions of years in the making, but more modestly they trace back to the Space Telescope Science Institute (STScI) in 1985. At the time the Hubble Space Telescope was still five years away from launching on a space shuttle. But Garth Illingworth, then deputy director of the STScI, was surprised one day when his boss, then director Riccardo Giacconi, who died in 2018, asked him to start thinking about what would come after Hubble much farther down the road. "I protested, saying we've got more than enough to do on Hubble," Illingworth recalls. But Giacconi was insistent: "Trust me, it'll take a long time," he said. So Illingworth and a handful of others got to work, drawing up concept ideas for what became known as the Next Generation Space Telescope (NGST), later renamed JWST after a former NASA administrator.

Hubble would be transformational, but astronomers knew its capabilities would be limited by its observations in visible light. As light from a very distant galaxy travels across the cosmic abyss, it is stretched by the expansion of the universe–a broadening of wavelengths known as redshift. The higher the redshift value, the more stretching the light has experienced, and thus the more distant its source galaxy. Redshifts for early galaxies are so high that their emitted visible light has stretched into infrared by the time it arrives at our telescopes, which is why Hubble could not see them. The NGST, in contrast, would observe in infrared and would boast a very large (and very cold) starlight-gathering mirror, allowing it to peer much deeper into the universe. "Everybody realized that Webb would be the telescope for looking at early galaxies," Illingworth says. "That became the primary science goal."

The need for the telescope was highlighted in December 1995, when astronomers pointed Hubble at a seemingly empty patch of the sky for 10 consecutive days. Many experts predicted the extended observation would be a waste of resources, revealing at best a handful of dim galaxies, but instead the effort was richly rewarded. The resulting image, the Hubble Deep Field, showed the "empty" spot was filled with galaxies by the thousands, stretching back 12 billion years into the 13.8-billion-year history of our universe. "There were galaxies everywhere," says Illingworth, now an astrophysicist at the University of California, Santa Cruz. The Hubble Deep Field showed that the early universe was even more crowded and exciting than most anyone had expected, offering observational treasures to those who took the time and care to properly look. Yet, impressive as Hubble's Deep Field was, astronomers wanted more.

After more than two decades of labor at a cost of some $10 billion, JWST finally launched on Christmas Day 2021. The telescope reached its deep-space destination a month later, where it would endure exhaustive testing to ensure its optimal performance. By July 2022 it was ready to begin its long-awaited first year of science observations, known as Cycle 1. Part of the telescope's early time was devoted to high-impact programs across a range of disciplines from which data would immediately be made public. Two of those, CEERS (the Cosmic Evolution Early Release Science Survey) and GLASS (the Grism Lens–Amplified Survey from Space), independently spent dozens of hours looking for galaxies in the early universe by staring at separate small portions of the sky. Not much was expected—perhaps a slightly more ornate version of the Hubble Deep Field but nothing more. Steven Finkelstein of the University of Texas at Austin, the lead on CEERS, says extremely distant galaxies were predicted to pop up only "after a few cycles of data" from multiple programs.

Instead, much to the surprise of astronomers, extremely distant galaxies came into view immediately. Hubble's record for the most distant known galaxy had been GN-z11, spotted in 2015 at a redshift of 11 thanks to a 2009 upgrade that enhanced the telescope's modest infrared capabilities. A redshift of 11 corresponds to a cosmic age of about 400 million years, a point at the brink of when galaxy formation was thought to begin. But from the very first GLASS data, two teams—one led by Naidu in that breathless late-night discovery—independently found GLASS-z13 at a redshift of 13, some 70 million years farther back in time.

In their quest for quick results, the researchers relied on redshift estimates derived from simple brightness-based measurements. These are easier to obtain but less precise than direct measurements of redshift, which require more dedicated observation time. Nevertheless, the simplified technique can be accurate, and here it suggested a galaxy that was unexpectedly bright and big, already bearing a mass of stars equivalent to a billion suns, just a few hundred times less than that of the Milky Way's stellar population, despite our own galaxy being billions of years more mature. "This was beyond our most optimistic expectations," says Tommaso Treu, an astronomer at the University of California, Los Angeles, and the lead on GLASS.

The record didn't last long. In the following days, dozens of galaxy candidates from CEERS and GLASS sprang into view with estimated redshifts as high as 20—just 180 million years after the big bang—some with disklike structures that were not expected to manifest so early in cosmic history. Another team, meanwhile, found evidence for galaxies the size of our Milky Way at a redshift of 10, less than 500 million years after the big bang.

Such behemoths emerging so rapidly defies expectations set by cosmologists' standard model of the universe's evolution. Called Lambda CDM (LCDM), this model incorporates scientists' best estimates for the properties of dark energy and dark matter, which collectively act to dominate the emergence of large-scale cosmic structures. ("Lambda" refers to dark energy, and "CDM" refers to dark matter that is relatively sluggish, or "cold.") "Even if you took everything that was available to form stars and snapped your fingers instantaneously, you still wouldn't be able to get that big that early," says Michael Boylan-Kolchin, a cosmologist at the University of Texas at Austin. "It would be a real revolution."

How to Build a Galaxy

To understand the dilemma, a brief refresher is needed. In the first second after the big bang, our universe was an almost inconceivably hot and dense soup of primordial particles. Over the next three minutes, as the cosmos expanded and cooled, the nuclei of helium and other very light elements began to form. Fast-forward 400,000 years, and the universe was cold enough for the first atoms to appear. When the universe was about 100 million years old, theorists say, conditions were finally right for the emergence of the first stars. These giant fireballs of mostly hydrogen and helium were uncontaminated by heavier elements found in modern-day stars, so they possessed significantly different properties. Larger and brighter than today's stars, these first suns coalesced in protogalaxies—clusters of gas that clung to vast, invisible scaffolds of dark matter. Gravity guided the subsequent interactions between these protogalaxies, which eventually merged to form larger galaxies. This process of becoming, of the early universe's chaos giving way to the more orderly cosmos we know today, is thought to have taken about a billion years.

JWST's discovery of bright galaxies in the early cosmos challenges this model. "We should see lots of these little protogalactic fragments that have not yet merged to make a big galaxy," says Stacy McGaugh, a cosmologist at Case Western Reserve University. "Instead we're seeing a few things that are already big galaxies." Some of these galaxies may be impostors, much closer galaxies shrouded in dust that makes them look dimmer and farther away when brightness-based measurements are used. Follow-up observations of GLASS-z13 in August 2022 by the Atacama Large Millimeter Array (ALMA) in Chile, however, suggested that is not the case for this candidate, because ALMA did not see evidence for large amounts of dust. "I think we can exclude low-redshift interlopers," says Tom Bakx, an astronomer at Nagoya University in Japan, who led the observations. Yet the lack of dust means ALMA struggled to see the galaxy at all, showing how difficult it could be for telescopes to confirm observations made using JWST's advanced capabilities. "The good news is there's nothing detected," Naidu says. "The bad news is there's nothing detected." Only JWST, in this case, can follow up itself.

The most startling potential explanation is that the canonical LCDM cosmological model is wrong and requires revision. "These results are very surprising and hard to get in our standard model of cosmology," Boylan-Kolchin says. "And it's probably not a small change. We'd have to go back to the drawing board." One controversial idea is modified Newtonian dynamics (MOND), which posits that dark matter does not exist and that effects attributed to it can instead be explained by large-scale fluctuations in gravity. JWST's observations could support such a theory. "MOND has had a lot of its predictions come true; this is another one of them," says McGaugh, who is one of the idea's leading proponents. Others remain unconvinced. "So far everything that we've tried to test MOND hasn't been able to really provide a satisfactory answer," says Jeyhan Kartaltepe, an astrophysicist at the Rochester Institute of Technology.

One simpler possible solution is that galaxies in the early universe might have little or no dust, making them appear brighter. This scenario could confound efforts to calculate the galaxies' true masses and could perhaps also explain ALMA's difficulty spotting GLASS-z13. "It could be that supernovae didn't have enough time to produce the dust, or maybe in the initial phases [of galaxy formation] the dust is expelled from galaxies," says Andrea Ferrara, an astronomer at the Scuola Normale Superiore in Italy, who has proposed such a possibility. Mason and her colleagues suggest that, alternatively, in its observations of the early universe JWST may so far be seeing only the very brightest young galaxies, which should be the easiest to spot. "Maybe there's something happening in the early universe that means it's easier for some galaxies to form stars," she says.

David Spergel, a theoretical astrophysicist and current president of the Simons Foundation in New York City, agrees. "I think what we're seeing is that high-mass star formation is very efficient in the early universe," he says. "The gas pressures are higher. The temperatures are higher. That has an enormous impact on the environment for star formation." Magnetic fields might have arisen earlier in the universe than we thought, driving material to kick-start the birth of stars. "We might be seeing a signature of magnetic fields emerging very early in the universe's history," Spergel says.

A Rush to Break the Universe

The rapid flow of scientific papers from JWST's initial observations is no fluke; when the first data started streaming down, astronomers were eagerly waiting. "People had been working on their pipelines for years," Boylan-Kolchin says. Instead of the traditional peer-review processes, which can take months, astronomers published on arXiv, a website where scientific papers can be uploaded after minimal review by moderators but well before formal peer review. A new form of peer review unfolded in near real time on Twitter and other social media platforms. "It's science by arXiv," Naidu says. The resulting frenzy was intense–and surprising. "I expected a lot of activity," says Nancy Levenson, STScI's interim director. "But I underestimated the amount."

The result was that findings could be quickly publicized and discussed, but some fear at a cost. "People were rushing things a little bit," says Klaus Pontoppidan, JWST's project scientist at STScI. "The gold standard is a refereed, peer-reviewed paper." Early calibration issues with JWST, for example, may have affected some results. Nathan Adams of the University of Manchester in England and his colleagues found there could be dramatic changes, with one galaxy identified at a redshift of 20.4 recalibrated to just 0.7. "We need to calm down a little bit," Adams says. "It's a bit too early to say we've completely broken the universe."

Such issues are unlikely to eradicate all of JWST's high-redshift galaxies, however, given their sheer number. "It's more likely that the early universe is different from what we predicted," Finkelstein says. "The odds are small that we're all wrong." Astronomers are now racing to conduct follow-up observations with JWST. Levenson says she's currently reviewing about a dozen proposals from various groups asking for additional JWST observing time, most of which are seeking to scrutinize high-redshift galaxy candidates. "Considering the excitement and importance of these early discoveries, we thought it was appropriate to ask for a little bit of time to confirm them," says Treu, who put forward one of the proposals.

Other programs are now set to hunt for distant galaxies, such as COSMOS-Web, led by Kartaltepe, which is expected to hugely increase the known population of early galaxies by observing a wider swath of sky for hundreds of hours. "We estimate there are thousands we'll be able to detect," she says. Future projects might look for evidence of those first protogalaxies, perhaps using the explosive deaths of supersized first stars in especially luminous and energetic supernovae as markers for their existence. Some estimates suggest JWST could see as far as a redshift of 26, just 120 million years after the big bang, a cosmic blink of an eye. Much other work will be done to follow up the growing list of high-redshift candidates. "Even confirming a handful of these would be quite amazing," Naidu says. "It would demonstrate we're not getting fooled."

JWST has ushered in a new era of science, and despite the uncertainties, the rapid communication of new discoveries has invigorated astronomers. "It's been fantastic," Treu says. "It's really wonderful to see the community so engaged and excited." Now the question is, if we can truly believe what we are seeing, is it time to reappraise our understanding of the dawn of time? "We're peering into the unknown," Mason says.

About the Author

Jonathan O'Callaghan is an award-winning freelance journalist covering astronomy, astrophysics, commercial spaceflight and space exploration.

Section 2: Black Holes

Black Hole Discovery Helps to Explain Quantum Nature of the Cosmos

By Edgar Shaghoulian

Where did the universe come from? Where is it headed? Answering these questions requires that we understand physics on two vastly different scales: the cosmological, referring to the realm of galaxy superclusters and the cosmos as a whole, and the quantum—the counterintuitive world of atoms and nuclei.

For much of what we would like to know about the universe, classical cosmology is enough. This field is governed by gravity as dictated by Einstein's general theory of relativity, which doesn't concern itself with atoms and nuclei. But there are special moments in the lifetime of our universe—such as its infancy, when the whole cosmos was the size of an atom—for which this disregard for small-scale physics fails us. To understand these eras, we need a quantum theory of gravity that can describe both the electron circling an atom and Earth moving around the sun. The goal of quantum cosmology is to devise and apply a quantum theory of gravity to the entire universe.

Quantum cosmology is not for the faint of heart. It is the Wild West of theoretical physics, with little more than a handful of observational facts and clues to guide us. Its scope and difficulty have called out to young and ambitious physicists like mythological sirens, only to leave them foundering. But there is a palpable feeling that this time is different and that recent breakthroughs from black hole physics—which also required understanding a regime where quantum mechanics and gravity are equally important—could help us extract some answers in quantum cosmology. The fresh optimism was clear at a recent virtual physics conference I attended, which had a dedicated discussion session about the crossover between the two fields. I expected this event to be sparsely attended, but instead many of the luminaries in physics were there, bursting with ideas and ready to get to work.

Event Horizons

The first indication that there is any relation between black holes and our universe as a whole is that both manifest "event horizons"—points of no return beyond which two people seemingly fall out of contact forever. A black hole attracts so strongly that at some point even light—the fastest thing in the universe—cannot escape its pull. The boundary where light becomes trapped is thus a spherical event horizon around the center of the black hole.

Our universe, too, has an event horizon—a fact confirmed by the stunning and unexpected discovery in 1998 that not only is space expanding, but its expansion is *accelerating*. Whatever is causing this speedup has been called dark energy. The acceleration traps light just as black holes do: as the cosmos expands, regions of space repel one another so strongly that at some point not even light can overcome the separation. This inside-out situation leads to a spherical cosmological event horizon that surrounds us, leaving everything beyond a certain distance in darkness. There is a crucial difference between cosmological and black hole event horizons, however. In a black hole, spacetime is collapsing toward a single point—the singularity. In the universe at large, all of space is uniformly growing, like the surface of a balloon that is being inflated. This means that creatures in faraway galaxies will have their own distinct spherical event horizons, which surround them instead of us. Our current cosmological event horizon is about 16 billion light-years away. As long as this acceleration continues, any light emitted today that is beyond that distance will never reach us. (Cosmologists also speak of a particle horizon, which confusingly is often called a cosmological horizon as well. This refers to the distance beyond which light emitted in the early universe has not yet had time to reach us here on Earth. In our tale, we will be concerned only with the cosmological event horizon, which we will often just call the cosmological horizon. These are unique to universes that accelerate, like ours.)

The similarities between black holes and our universe don't end there. In 1974 Stephen Hawking showed that black holes are not completely black: because of quantum mechanics, they have a temperature and therefore emit matter and radiation, just as all thermal bodies do. This emission, called Hawking radiation, is what causes black holes to eventually evaporate away. It turns out that cosmological horizons also have a temperature and emit matter and radiation because of a very similar effect. But because cosmological horizons surround us and the radiation falls inward, they reabsorb their own emissions and therefore do not evaporate away like black holes.

Hawking's revelation posed a serious problem: if black holes can disappear, so can the information contained within them—which is against the rules of quantum mechanics. This is known as the black hole information paradox, and it is a deep puzzle complicating the quest to combine quantum mechanics and gravity. But in 2019 scientists made dramatic progress. Through a confluence of conceptual and technical advances, physicists argued that the information inside a black hole can actually be accessed from the Hawking radiation that leaves the black hole.

This discovery has reinvigorated those of us studying quantum cosmology. Because of the mathematical similarities between black holes and cosmological horizons, many of us have long believed that we couldn't understand the latter without understanding the former. Figuring out black holes became a warm-up problem—one of the hardest of all time. We haven't fully solved our warm-up problem yet, but now we have a new set of technical tools that provide beautiful insight into the interplay of gravity and quantum mechanics in the presence of black hole event horizons.

Entropy and the Holographic Principle

Part of the recent progress on the black hole information paradox grew out of an idea called the holographic principle, put forward in the 1990s by Gerard 't Hooft of Utrecht University in the Netherlands and Leonard Susskind of Stanford University. The holographic principle states that a theory of quantum gravity that can describe black holes should be formulated not in the ordinary three spatial dimensions that all other physical theories use but instead in two dimensions of space, like a flat piece of paper. The primary argument for this approach is quite simple: a black hole has an entropy—a measure of how much stuff you can stick inside it—that is proportional to the two-dimensional area of its event horizon.

Contrast this with the entropy of a more traditional system—say, a gas in a box. In this case, the entropy is proportional to the three-dimensional volume of the box, not the area. This is natural: you can stick something at every point in space inside the box, so if the volume grows, the entropy grows. But because of the curvature of space within black holes, you can actually increase the volume without affecting the area of the horizon, and this will not affect the entropy! Even though it naively seems you have three dimensions of space to stick stuff in, the black hole entropy formula tells you that you have only two dimensions of space, an area's worth. So the holographic principle says that because of the presence of black holes, quantum gravity should be formulated as a more prosaic nongravitational quantum system in fewer dimensions. At least then the entropies will match.

The idea that space might not be truly three-dimensional is rather compelling, philosophically. At least one dimension of it might be an emergent phenomenon that arises from its deeper nature rather than being explicitly hardwired into the fundamental laws. Physicists who study space now understand that it can emerge from a large collection of simple constituents, similar to other emergent phenomena such as consciousness, which seems to arise from basic neurons and other biological systems.

One of the most exciting aspects of the progress in the black hole information paradox is that it points toward a more general understanding of the holographic principle, which previously had been made precise only in situations very different from our real universe. In the calculations from 2019, however, the way the information inside of the black hole is encoded in the Hawking radiation is mathematically analogous to how a gravitational system is encoded in a lower-dimensional nongravitational system according to the holographic principle. And these techniques can be used in situations more like our universe, giving a potential avenue for understanding the holographic principle in the real world. A remarkable fact about cosmological horizons is that they also have an entropy, given by the exact same formula as the one we use for black holes. The physical interpretation of this entropy is much less clear, and many of us hope that applying the new techniques to our universe will shed light on this mystery. If the entropy is measuring how much stuff you can stick beyond the horizon, as with black holes, then we will have a sharp bound on how much stuff there can be in our universe.

Outside Observers

The recent progress on the black hole information paradox suggests that if we collect all the radiation from a black hole as it evaporates, we can access the information that fell inside the black hole. One of the most important conceptual questions in cosmology is whether the same is possible with cosmological event horizons. We think they radiate like black holes, so can we access what is beyond our cosmological event horizon by collecting its radiation? Or is there some other way to reach across the horizon? If not, then most of our vast, rich universe will eventually be lost forever. This is a grim image of our future—we will be left in the dark.

Almost all attempts to get a handle on this question have required physicists to artificially extricate themselves from the accelerating universe and imagine viewing it from the outside. This is a crucial simplifying assumption, and it more closely mimics a black hole, where we can cleanly separate the observer from the system simply by placing the observer far away. But there seems to be no escaping our cosmological horizon; it surrounds us, and it moves if we move, making this problem much more difficult. Yet if we want to apply our new tools from the study of black holes to the problems of cosmology, we must find a way to look at the cosmic horizon from the outside.

There are different ways to construct an outsider view. One of the simplest is to consider a hypothetical auxiliary universe that is quantum-mechanically entangled with our own and investigate whether an observer in the auxiliary universe can access the information in our cosmos, which is beyond the observer's horizon. In work I did with Thomas Hartman and Yikun Jiang, both at Cornell University, we constructed examples of auxiliary universes and other scenarios and showed that the observer can access information beyond the cosmological horizon in the same way that we can access information beyond the black hole horizon. (A complementary paper by Yiming Chen of Princeton University, Victor Gorbenko of EPFL in Switzerland and Juan Maldacena of the Institute for Advanced Study [IAS] in Princeton, N.J., showed similar results.)

But these analyses all have one serious deficiency: when we investigated "our" universe, we used a model universe that is contracting instead of expanding. Such universes are much simpler to describe in the context of quantum cosmology. We don't completely understand why, but it's related to the fact that we can think of the interior of a black hole as a contracting universe where everything is getting squished together. In this way, our newfound understanding of black holes could easily help us study this type of universe

Even in these simplified situations, we are puzzling our way through some confusing issues. One problem is that it's easy to construct multiple simultaneous outsider views so that each outsider can access the information in the contracting universe. But this means multiple people can reach the same piece of information and manipulate it independently. Quantum mechanics, however, is exacting: not only does it forbid information from being destroyed, it also forbids information from being replicated. This idea is known as the no-cloning theorem, and the multiple outsiders seem to violate it. In a black hole, this isn't a problem, because although there can still be many outsiders, it turns out that no two of them can independently access the same piece of information in the interior. This limit is related to the fact that there is only one black hole and therefore just one event horizon. But in an expanding spacetime, different observers have different horizons. Recent work that Adam Levine of the Massachusetts Institute of Technology and I did together, however, suggests that the same technical tools from the black hole context work to avoid this inconsistency as well.

Toward a Truer Theory

Although there has been exciting progress, so far we have not been able to directly apply what we learned about black hole horizons to the cosmological horizon in our universe because of the differences between these two types of horizons.

The ultimate goal? No outsider view, no contracting universe, no work-arounds: we want a complete quantum theory of our expanding universe, described from our vantage point within the belly of the beast. Many physicists believe our best bet is to come up with a holographic description, meaning one using fewer than the usual three dimensions of space. There are two ways we can do this. The first is to use tools from string theory, which treats the elementary particles of nature as vibrating strings. When we configure this theory in exactly the right way, we can provide a holographic description of certain black hole horizons. We hope to do the same for the cosmological horizon. Many physicists have put a lot of work into this approach, but it has not yet yielded a complete model for an expanding universe like ours.

The other way to divine a holographic description is by looking for clues from the properties that such a description should have. This approach is part of the standard practice of science—use data to construct a theory that reproduces the data and hope it makes novel predictions as well. In this case, however, the data themselves are also theoretical. They are things we can reliably calculate even without a complete understanding of the full theory, just as we can calculate the trajectory of a baseball without using quantum mechanics. The idea works as follows: we calculate various things in classical cosmology, maybe with a little bit of quantum mechanics sprinkled in, but we try to avoid situations where quantum mechanics and gravity are equally important. This forms our theoretical data. For example, Hawking radiation is a piece of theoretical data. And what must be true is that the full, exact theory of quantum cosmology should be able to reproduce this theoretical datum in an appropriate regime, just as quantum mechanics can reproduce the trajectory of a baseball (albeit in a much more complicated way than classical mechanics).

Leading the charge in extracting these theoretical data is a powerful physicist with a preternatural focus on the problems of quantum cosmology: Dionysios Anninos of King's College London has been working on the subject for more than a decade and has provided many clues toward a holographic description. Others around the world have also joined the effort, including Edward Witten of IAS, a figure who has towered over quantum gravity and string theory for decades but who tends to avoid the Wild West of quantum cosmology. With his collaborators Venkatesa Chandrasekaran of IAS, Roberto Longo of the University of Rome Tor Vergata and Geoffrey Penington of the University of California, Berkeley, he is investigating how the inextricable link between an observer and the cosmological horizon affects the mathematical description of quantum cosmology.

Sometimes we are ambitious and try to calculate theoretical data when quantum mechanics and gravity are equally important. Inevitably we have to impose some rule or guess about the behavior of the full, exact theory in such instances. Many of us believe that one of the most important pieces of theoretical data is the amount and pattern of entanglement between constituents of the theory of quantum cosmology. Susskind and I formulated distinct proposals for how to compute these data, and in hundreds of e-mails exchanged during the pandemic, we argued incessantly over which was more reasonable. Earlier work by Eva Silverstein of Stanford, another brilliant physicist with a long track record in quantum cosmology, and her collaborators provides yet another proposal for computing these theoretical data.

The nature of entanglement in quantum cosmology is a work in progress, but it seems clear that nailing it will be an important step toward a holographic description. Such a concrete, calculable theory is what the subject desperately needs, so that we can compare its outputs with the wealth of theoretical data that are accumulating from scientists. Without this theory, we will be stuck at a stage akin to filling out the periodic table of elements without the aid of quantum mechanics to explain its patterns.

There is a rich history of physicists quickly turning to cosmology after learning something novel about black holes. The story has often been the same: we've been defeated and humbled, but after licking our wounds, we've returned to learn more from what black holes have to teach us. In this instance, the depth of what we've realized about black holes and the breadth of interest in quantum cosmology from scientists around the world may tell a different tale.

About the Author

Edgar Shaghoulian is a theoretical physicist now at the University of California, Santa Cruz. His work focuses on black holes and quantum cosmology.

Colliding Supermassive Black Holes Discovered in Nearby Galaxy

By Allison Parshall

Our galaxy, the Milky Way, is on a collision course. Some four billion years from now, the Milky Way and its large neighboring galaxy, Andromeda, will begin a spiraling gravitational dance, merging over hundreds of millions of years to form one larger object.

Such galaxy mergers are happening all the time, all over the universe, and they are important building blocks for larger-scale cosmic structure. Astrophysicists have many questions about this extraordinary process, but one mystery has proved especially captivating: When large galaxies merge, what happens to the supermassive black holes that decades of observations have revealed to be lurking at their centers?

Logically, these giant black holes—each millions to billions of times heavier than our sun—must collide and merge, too. Such mergers can channel huge volumes of material into the black holes, sparking violent astrophysical outbursts that shape star formation and other processes in their host galaxies. But astronomers have so far only seen snapshots of this long process, from when the black holes are still tens to hundreds of light-years apart. The closer the merging black holes get, the harder they become to distinguish from each other, blurring the picture for theorists seeking to understand how this process works.

Now an international team of scientists has announced the discovery of two active supermassive black holes close to Earth in a new study in the *Astrophysical Journal Letters*. At an estimated 125 million and 200 million times the mass of the sun, respectively, these black holes sit about 500 million light-years away from us, gobbling up gas and dust at the center of UGC 4211, a galaxy that is still reeling from a merger.

"This pair was really exciting because they're so close to each other and they're so nearby," says Chiara Mingarelli, one of the study's authors and an astrophysicist at the Flatiron Institute in New York City and the University of Connecticut. Separated by only some 750 light-years, or 230 parsecs, they are the closest pair of black holes that scientists have been able to confirm by measuring multiple wavelengths of light. These black holes' proximity to Earth and each other may provide a unique opportunity for fundamental studies of giant black hole mergers, as well as one of their most elusive by-products: ripples in spacetime called gravitational waves.

The project began nearly 10 years ago, when astrophysicist Michael Koss started using some of the world's largest telescopes to search the sky for "active" pairs of supermassive black holes–that is, those that are feeding and burping out blasts of intense radiation. After examining nearly 100 objects, he found six that were actually hidden merging pairs. Of these, UGC 4211's stood out as being much closer together than the rest.

"You can see it. I mean, literally, you can see on the [images] there are two sources," says Koss, who works at Eureka Scientific, an astrophysics research institute in California. The high resolution was possible, in part, because of UGC 4211's proximity to Earth.

This paper "has really pushed the limit" of what is possible to observe, says Zoltan Haiman, an astronomer at Columbia University, who was not involved in the study. Astronomers have observed one pair of binary black holes in even closer proximity to each other before but only with radio telescopes. In the new study, the team confirmed its findings using multiwavelength data from several different telescopes–an important step because the field has had false positives in the past, Mingarelli says. "You can only trick the eye in so many wavelengths," she says. The team studied the system using the W. M. Keck Observatory in Hawaii, the Very Large Telescope and the Atacama Large Millimeter/submillimeter Array in Chile, and the Hubble Space Telescope in orbit around Earth.

UGC 4211's two black holes are thought to be midway through their merger. As their extended cosmic duet progresses, the pair will draw even closer together as surrounding swarms of stars and gas siphon away their orbital momentum. The authors predict that the dance will end in approximately 200 million years, when the two supermassive black holes at last fully merge to become one.

"This seems to be right smack in line with a lot of our paradigms, so that's a good thing. It's not breaking astrophysics," says Jeremy Schnittman, an astrophysicist at NASA's Goddard Space Flight Center, who was not involved in the recent study.

At least, that's the case for now. Much remains unknown about the mechanics of merging monstrous black holes, especially during later stages, when the black holes approach each other so closely that they cannot be clearly distinguished. "What happens once they're even closer than this, down at parsecs or less, is still a really big question," says Sarah Burke-Spolaor, an astronomer at West Virginia University, who was not involved in the new results.

Like many of her peers, Burke-Spolaor is especially fascinated by each merger's final phase, when the black holes spiral together so violently that they shake the fabric of spacetime itself, producing copious gravitational waves. After first detecting such emissions in 2015–a discovery that netted a Nobel Prize in Physics–astronomers now routinely study these last gasps from merging black holes using specialized observatories that are as different from light-gathering telescopes as ears are from eyes. Most of those studies, however, concern pairs of black holes that are far smaller than their supermassive kin. Tuning in to the immense gravitational waves from the universe's largest merging black holes requires a new generation of even more advanced observatories that scientists and engineers have scarcely begun to build.

The payoff should be worthwhile: Such huge mergers are thought to be the most common contributor to the "gravitational wave background," the as-yet-undetected totality of spacetime ripples from sources scattered across the entire observable universe, imprinted across the entire sky. Detecting and mapping this background, Mingarelli says, would yield "the cosmic merger history of supermassive black holes" and a wealth of other cosmologically vital information. And studying the fine details of merging systems like UGC 4211 can help researchers better understand what they're seeing if or when they finally manage to glimpse the universe's gravitational wave background.

Koss hopes that future studies of UGC 4211 will be especially fruitful, given its proximity to Earth. This system is easier to observe from our place in the universe, "just like you can see the leaves on the nearby trees in the forest, not distant ones," he says.

The discovery also suggests that there may be more merging black holes to observe than we previously thought because Koss found it from a sample of fewer than 100 active black holes. "There's always a chance that the study just got lucky," Burke-Spolaor says. But even if it was a fluke, scientists will continue searching for more examples to fill in the final steps of this cosmic choreography.

About the Author

Allison Parshall is a science journalist, multimedia editor, podcast host and former news intern at Scientific American.

Did Astronomers Just Discover Black Holes from the Big Bang?

By Nola Taylor Tillman

In the nearly five years since their first direct detection, gravitational waves have become one of the hottest topics in astronomy. With facilities such as the Laser Interferometer Gravitational-Wave Observatory (LIGO), researchers have mostly used these ripples in spacetime to study the inner workings of merging black holes, but LIGO has also detected gravitational waves from other sorts of celestial crashes, such as the collisions of ultradense stellar remnants called neutron stars. Sometimes, however, LIGO serves up gravitational waves that leave astronomers scratching their heads–as was the case for GW190425, an event detected last April that was recently attributed to a neutron star merger.

The trouble is that LIGO's data suggest this neutron star pair was substantially overweight–collectively, some 3.4 times the mass of the sun, which is half a solar mass heavier than the most massive neutron star binaries ever seen. "It is the heaviest known by a pretty wide margin," says Chad Hanna, an astrophysicist at Pennsylvania State University who hunts gravitational waves.

That extra weight has some theorists suspecting that GW190425 did not arise from colliding neutron stars at all but rather something much more exotic: A merger of two primordial black holes (PBHs), never before seen objects that are considered a dark horse candidate for dark matter–the invisible, unidentified *something* that makes up most of the matter in the universe. Theorized to have formed from density fluctuations in the very early universe, these ancient black holes could still exist today and could explain the mass discrepancy identified in the recent LIGO observations.

Almost a half-century ago, cosmologist Stephen Hawking proposed that PBHs could have sprung fully formed from regions of the infant universe that were especially dense with matter. Since then, the idea's popularity among astrophysicists and cosmologists has wildly waxed and waned. Today, in the absence of direct evidence for their existence, PBHs are seen by many researchers as a hypothesis of last resort, only to be considered when no other scenario readily fits observations. The possibility that PBHs are real and widespread throughout the universe cannot yet be dismissed, however—especially as searches for other dark matter candidates come up empty.

PBHs make an appealing candidate for dark matter for several reasons, but the most important one is that, being black holes, they are quite dark yet still pack a hefty gravitational pull. Despite that fact, Hanna says that if PBHs were abundant enough to account for all of the universe's dark matter, astronomical surveys that hunted for them should not have come up empty. Consequently, he adds, PBHs can only make up a small fraction of dark matter—if they exist at all.

Not everyone agrees. "Primordial black holes can comprise the whole of dark matter," says Juan García-Bellido, a theoretical cosmologist at the Autonomous University of Madrid. The trick, he adds, is for the ancient objects to exhibit an array of masses rather than a single definitive size. If PBHs run the gamut from a thousand times less massive than the sun to a billion times larger, they could make up *all* of the universe's dark matter. "All published constraints that claim to rule out primordial black holes as dark matter assume they exist in a monochromatic, or single-mass, spectrum and are uniformly distributed in space," García-Bellido says. For such large mass ranges to manifest, the PBHs would have to cluster in compact groups in which they could occasionally collide, merge and grow larger.

Because PBHs would have been created shortly after the big bang, they initially could have easily connected with one another. The early universe was a much smaller place than it is today after dramatically expanding for nearly 14 billion years, making it easier for the objects to find other PBHs and pair up with them. As the universe continued to expand, and the first stars and galaxies emerged, however, those connections would have become increasingly rare. So while it is possible that LIGO has observed merging PBHs, it is unlikely, according to astronomer Katerina Chatziioannou, a LIGO team member at the Flatiron Institute in New York City and co-author of a study set to appear in the *Astrophysical Journal Letters* that pegs GW190425 as the product of colliding neutron stars.

Last April, alerted to LIGO's detection of GW190425, telescopes around the world hunted for a corresponding electromagnetic signal that would typically be expected from the explosive collision of two neutron stars. But the skies remained dark, as they would if a pair of primordial black holes had slammed together. "We would not have expected any light to have come out of the merger of two primordial black holes," Chatziioannou says. Even so, she adds, the lack of light does not rule out neutron stars. Massive neutron stars could have had a relatively placid merger, collapsing directly into a black hole before sparking any celestial fireworks. It is also possible that the event's location in the sky could be from somewhere that Earth-bound telescopes could not probe, such as in a region behind the sun. "There are good reasons for them to have missed it if there was any light," she says.

The newest observations provide only tantalizing hints that PBHs might be out there, occasionally coming together in the cosmic dark. A clearer signature would come from a pair of black holes in which each weighed less than the sun. "If you find a black hole below a solar mass, that, at the very least, comes from a mechanism that no one has predicted, astrophysically, outside of a primordial black hole," Hanna says. García-Bellido concurs. "The smoking gun would be the discovery of a less than one-solar-mass black hole—or a black hole with a mass larger than 50 [times the sun]," he says.

Although LIGO's observations could mark the first detection of PBHs, both Chatziioannou and Hanna agree that it is more likely that the gravitational waves merely came from overweight neutron stars. Theories for the formation of such bulky neutron stars already exist, and they require no speculative scenarios from the dawn of the universe. "It's definitely a lot less likely that [the sources of these events] are primordial black holes than just neutron stars that are heavier than what we see in the galaxy," Chatziioannou says. "It's not impossible; it's just less likely."

Although Hanna calls the case for GW190425 as a primordial black hole binary "weak," García-Bellido remains more optimistic. "*All* LIGO events could be due to primordial black holes," he says. Only time—and more data—will tell.

About the Author

Nola Taylor Tillman is a science writer with a focus on space and astronomy.

Astronomers Spy a Giant Runaway Black Hole's Starry Wake

By Kenna Hughes-Castleberry

What's invisible, weighs 20 million suns and zooms through space at more than 1,500 kilometers per second, leaving a long starry trail in its wake? If you guessed "a supermassive black hole that has escaped its host galaxy," you're probably right. At least that's the conclusion of an international team of researchers who found and studied the candidate runaway using some of the world's most powerful telescopes. The discovery is detailed in a paper published on April 6 in *Astrophysical Journal Letters*.

For reasons that are still poorly understood, supermassive black holes lurk at the centers of most large galaxies such as our own Milky Way. Despite their terrifying dimensions, these mysterious giants are usually quite inert and easy to overlook. They often only betray their presence with the celestial fireworks they create while swallowing immense volumes of gas and dust from their host galaxy.

All this makes supermassive black holes that have somehow "gone rogue" a strange and elusive breed. How do they escape their galactic host, and how can they be seen when they emit no light at all? Across the past decade, astronomers have only managed to identify a small number of other candidate nomads darkly drifting through the intergalactic depths—but none of these, the study authors say, are as convincing as their newfound escapee.

That black hole was discovered by chance when it first appeared as a faint linear streak in a Hubble Space Telescope observation of globular clusters. Such features are usually artifacts of cosmic rays striking Hubble's detectors, explains lead study author Pieter van Dokkum, an astronomer at Yale University. Further observations via the ground-based W. M. Keck Observatory on Mauna Kea in Hawaii revealed instead that the streak was a stream of young blue stars stretching across an astonishing 200,000 light-years.

That stellar stream is the clinching evidence, van Dokkum says. It's presumably a kind of cosmic contrail that arose from a feedstock of gas that the onrushing black hole shocked and compressed into stars. The stream traces back arrowlike to a probable source galaxy some 7.5 billion light-years from Earth that shows no signs of now harboring a giant, feasting black hole at its core. "It has long been hypothesized that every galaxy is embedded in a vast reservoir of hydrogen gas," van Dokkum says. "We think we're seeing the black hole lighting up that gas as it speeds away, giving us a rare view of this elusive material."

"From the ages of these stars," he adds, "we deduce that the black hole escaped about 40 million years ago and is barreling through space at the very high speed of about 1600 kilometers per second."

This speed is what excites many experts. "If this is a runaway black hole..., it is traveling *very* fast," says University of Cambridge astronomy professor Christopher Reynolds, who did not participate in the study. So fast, in fact, that any explanation that does not involve a behemoth black hole running roughshod between galaxies would seem unlikely.

Even so, astronomers hope further observations using NASA's James Webb Space Telescope and Chandra X-ray Observatory could eliminate any lingering doubts. "Much more data is needed to pinpoint this hypothesis," says study co-author Maria Luísa Buzzo, a Ph.D. candidate at the Swinburne University of Technology in Australia. "We need additional data to confirm this scenario and to decrease the uncertainties on the nature of the [stellar stream] feature," she says. Erin Bonning, an astronomer at Emory University, who was not involved with the work, agrees. "You'd want to see the direct gravitational effects of a compact, massive object," she says.

Such effects are crucial for explaining how something weighing millions or billions of suns can be kicked out of a galaxy in the first place. Astronomers have long known that big galaxies can grow from collisions and mergers of smaller ones. Giant galaxies can merge with one another, too, forcing the black holes at their respective centers into close proximity. "We don't know very well what happens next," van Dokkum says. "Over 50 years ago it was proposed that sometimes these black holes could be dislodged from the center in the aftermath and even flung out of galaxies altogether. This observation seems to confirm that idea, giving us insight into what happens when arguably the most awesome objects in the universe come together."

If two black holes become locked in a gravitational dance and then a third crashes in, the resulting instability can hurl one of the trio away at sufficient speed to exit the host galaxy entirely. Another possibility is that, rather than three black holes scattering off one another like billiards, two coming together as one could suffice. "The second channel involves two black holes merging and emitting gravitational [waves]," says Avi Loeb, an astrophysicist at Harvard University, who was not involved with the paper. "When they plunge toward each other, the [gravitational waves] will have a preferred direction in the common case when the two black holes do not have identical masses. As a result, the merger remnant will recoil in the opposite direction to conserve momentum."

How exactly supermassive black holes manage to reach intergalactic space may have important implications for the host galaxies they leave behind. For instance, the black holes may suppress—or promote—further bouts of star formation. Finding more of them whizzing across the cosmos could thus give researchers a powerful new tool to learn more about how galaxies are born, grow and even eventually die. The curious specifics of this latest candidate—serendipitously spied as a starry "scratch" on a Hubble image—lead Bonning and other experts to suspect that evidence for other similar escapees may already exist, as-yet-unnoticed and awaiting discovery in other data sets.

"If this is actually what's happening, and if this is common enough, you can just go and systematically search for this kind of thing," Bonning says. Augmented by machine learning to pick out the telltale streaks, future panoramic surveys from upcoming facilities such as NASA's Nancy Grace Roman Space Telescope and the ground-based Vera C. Rubin Observatory could perhaps reveal a treasure trove of these objects—each, in Bonning's words, "a needle in a haystack that you can't see."

About the Author

Kenna Hughes-Castleberry is science communicator at JILA (a joint physics research institute between the National Institute of Standards and Technology and the University of Colorado Boulder) and a freelance science journalist. Her main writing focuses are quantum physics, quantum technology, deep technology, social media and the diversity of people in these fields, particularly women and people from minority ethnic and racial groups.

Black Holes, Quantum Entanglement and the No-Go Theorem

By Zoë Holmes, Andrew Sornborger

Suppose someone–let's call her Alice–has a book of secrets she wants to destroy so she tosses it into a handy black hole. Given that black holes are nature's fastest scramblers, acting like giant garbage shredders, Alice's secrets must be pretty safe, right?

Now suppose her nemesis, Bob, has a quantum computer that's entangled with the black hole. (In entangled quantum systems, actions performed on one particle similarly affect their entangled partners, regardless of distance or even if some disappear into a black hole.)

A famous thought experiment by Patrick Hayden and John Preskill says Bob can observe a few particles of light that leak from the edges of a black hole. Then Bob can run those photons as qubits (the basic processing unit of quantum computing) through the gates of his quantum computer to reveal the particular physics that jumbled Alice's text. From that, he can reconstruct the book.

But not so fast.

Our recent work on quantum machine learning suggests Alice's book might be gone forever, after all.

Quantum Computers to Study Quantum Mechanics

Alice might never have the chance to hide her secrets in a black hole. Still, our new no-go theorem about information scrambling has real-world application to understanding random and chaotic systems in the rapidly expanding fields of quantum machine learning, quantum thermodynamics, and quantum information science.

Richard Feynman, one of the great physicists of the 20th century, launched the field of quantum computing in a 1981 speech, when he proposed developing quantum computers as the natural platform to simulate quantum systems. They are notoriously difficult to study otherwise.

Our team at Los Alamos National Laboratory, along with other collaborators, has focused on studying algorithms for quantum computers and, in particular, machine-learning algorithms–what some like to call artificial intelligence. The research sheds light on what sorts of algorithms will do real work on existing noisy, intermediate-scale quantum computers and on unresolved questions in quantum mechanics at large.

In particular, we have been studying the training of variational quantum algorithms. They set up a problem-solving landscape where the peaks represent the high-energy (undesirable) points of the system, or problem, and the valleys are the low-energy (desirable) values. To find the solution, the algorithm works its way through a mathematical landscape, examining its features one at a time. The answer lies in the deepest valley.

Entanglement Leads to Scrambling

We wondered if we could apply quantum machine learning to understand scrambling. This quantum phenomenon happens when entanglement grows in a system made of many particles or atoms. Think of the initial conditions of this system as a kind of information–Alice's book, for instance. As the entanglement among particles within the quantum system grows, the information spreads widely; this *scrambling* of information is key to understanding quantum chaos, quantum information science, random circuits and a range of other topics.

A black hole is the ultimate scrambler. By exploring it with a variational quantum algorithm on a theoretical quantum computer entangled with the black hole, we could probe the scalability and applicability of quantum machine learning. We could also learn something new about quantum systems generally. Our idea was to use a variational quantum algorithm that would exploit the leaked photons to learn about the dynamics of the black hole. The approach would be an optimization procedure—again, searching through the mathematical landscape to find the lowest point.

If we found it, we would reveal the dynamics inside the black hole. Bob could use that information to crack the scrambler's code and reconstruct Alice's book.

Now here's the rub. The Hayden-Preskill thought experiment assumes Bob can determine the black hole dynamics that are scrambling the information. Instead, we found that the very nature of scrambling prevents Bob from learning those dynamics.

Stalled Out on a Barren Plateau

Here's why: the algorithm stalled out on a barren plateau, which, in machine learning, is as grim as it sounds. During machine-learning training, a barren plateau represents a problem-solving space that is entirely flat as far as the algorithm can see. In this featureless landscape, the algorithm can't find the downward slope; there's no clear path to the energy minimum. The algorithm just spins its wheels, unable to learn anything new. It fails to find the solution.

Our resulting no-go theorem says that any quantum machine-learning strategy will encounter the dreaded barren plateau when applied to an unknown scrambling process.

The good news is, most physical processes are not as complex as black holes, and we often will have prior knowledge of their dynamics, so the no-go theorem doesn't condemn quantum machine learning. We just need to carefully pick the problems we apply it to. And we're not likely to need quantum machine learning to peer inside a black hole to learn about Alice's book—or anything else—anytime soon.

So, Alice can rest assured that her secrets are safe, after all.

This is an opinion and analysis article, and the views expressed by the author or authors are not necessarily those of Scientific American.

About the Author

Zoë Holmes is the Mark Kac Postdoctoral Fellow at Los Alamos National Laboratory whose research focuses on quantum algorithms.

Andrew Sornborger, a physicist, heads the Quantum Science Center and the Beyond Moore's Law program at Los Alamos National Laboratory.

Neutrinos from a Nearby Galaxy Reveal Black Hole Secrets

By Phil Plait

In the zoo of subatomic particles, neutrinos are strange beasts. Unlike more familiar particles such as electrons and protons, ghostly neutrinos barely interact with other matter at all: they can fly right through a planet as if it weren't even there. This makes it irritatingly difficult to detect them and, for neutrinos streaming in from cosmic objects in the sky, even harder to know exactly where they come from. In a recent study published in *Science*, however, researchers identified an extragalactic source for these subatomic particles.

For the first time, astronomers have confidently detected neutrinos from NGC 1068, a galaxy with a huge and actively feeding black hole in its center. The neutrinos are being created outside the black hole's "point of no return"–its event horizon–although it's not clear just how; several mechanisms are plausible. Scientists are hoping this discovery will change how they understand not just NGC 1068 but *all* such galaxies. As a bonus, they think the finding may have revealed the source of a faint glow of neutrinos we see everywhere we look in the sky.

Material that falls toward a black hole first forms a flattened accretion disk orbiting around it. Friction heats this disk of matter to incredible temperatures, causing it to glow so brightly it outshines the entire host galaxy. We call such galaxies "active," and they are among the most luminous objects in the universe.

In the case of NGC 1068, detecting that brilliant light is difficult because thick clouds of opaque cosmic dust absorb essentially all of it, letting virtually no signal out. This is where neutrinos' most annoying property is an advantage to us: they can pass right through those dust clouds and fly out into space, eventually reaching Earth. Still, we're left with the problem of detecting them. How do you measure neutrinos when they pass unscathed through your detector? The good news is that to neutrinos, matter is only *mostly* permeable. Although it's extraordinarily rare, some do manage to interact with matter—but it takes a very special kind of observatory to see it.

Located almost exactly at Earth's South Pole, the IceCube Neutrino Observatory is just such a place, and it's not your standard astronomical facility. For one thing, it doesn't use a mirror to collect and focus light from cosmic objects as telescopes do; instead it has a series of relatively simple optical sensors hung along dozens of vertical strings, creating a 3-D array of more than 5,000 sensors that can detect the locations and times of flashes of light.

For another, it's buried under more than a kilometer of Antarctic ice. When a neutrino travels through the ice, it has some small chance of slamming into the nucleus of one of the oxygen or hydrogen atoms in that ice. But actual impacts are exceedingly uncommon: trillions of neutrinos pass through every cubic centimeter of matter on Earth every second, but measurable physical interactions with that matter may only happen days apart.

When they do occur, they create high-speed subatomic shrapnel—particles moving away from the nuclear collision site at just under the speed of light. These then plow through the ice as well. Here's the fun part: they actually travel faster than light can move through the ice. No laws of physics are being broken, though. The speed of light in a vacuum is the ultimate cosmic speed limit, but light moves more slowly when it travels through matter. Particles cannot move faster than light in a vacuum, but they can travel faster than light through matter. When they do, they create a kind of photonic boom, like the shock wave created when something travels through air faster than the speed of sound. These faster-than-light events manifest as bright flashes of blue light called Cherenkov radiation. They can be seen for some distance through the clear Antarctic ice and can be picked up by IceCube's detectors.

This phenomenon allows scientists to detect neutrino events from space, but there's a problem with unwanted events that mimic the desired signals. Subatomic particles from other sources in the universe called cosmic rays can hit our atmosphere and create similar flashes of light, confusing the measurements. Scientists can differentiate between the two kinds of signals in a clever way, though: by using Earth itself as an immense filter. Neutrinos coming from space will come from every direction, including up through Earth. Cosmic rays, however, will come only from the sky above the Antarctic observatory because they can't beam straight through Earth as neutrinos do. The detectors in IceCube can measure direction and filter out the events coming from above, thus ensuring scientists keep only the hits from cosmic neutrinos.

IceCube has detected millions of neutrinos overall, but only a few hundred at most appear to have come from bona fide cosmological objects. Some things out there in the universe are the sources of these neutrinos. The question is, What are they?

Looking over data taken from 2011 to 2020, the IceCube Collaboration–a huge collection of scientists, engineers, data analysts, and more–very carefully processed every event detected. Using the directional information from the flashes to trace the trajectories of the incoming cosmic neutrinos, they found several spots on the sky that appeared to be statistically significant sources of neutrinos.

The detection with the largest number of neutrinos? A total of 79 (plus or minus 20 or so) neutrinos over that period coming from the direction of NGC 1068.

This lovely spiral galaxy is relatively close–a mere 47 million light-years from us–and bright enough to be spotted with binoculars. Earlier work analyzing IceCube neutrinos pointed to NGC 1068 as a possible source, but the data weren't strong enough at the time to claim a discovery. These results change that.

The detection of neutrinos ostensibly coming from this active galaxy is a big deal. The neutrinos that the astronomers saw have phenomenally high energy: more than a tera-electron volt each. That is trillions of times the energy of the visible-light photons we see coming from the galaxy. The particles' huge energy must be created in an extremely powerful cosmic particle accelerator, and with an actively feeding big black hole, several options are possible.

For example, the turbulent ionized miasma of matter above and below the disk of material around the black hole is infernally hot and contains powerful magnetic fields that can pump vast energies into particles, accelerating them almost to light speed. Another way involves the magnetic field in that accretion disk getting twisted up near the black hole, creating twin vortices such as tornadoes, called jets, that can fling particles away at high speeds. Shock waves generated in the jets as charged particles slam into one another can also produce the energies needed for high-energy neutrinos. Such jets are known to exist in NGC 1068.

Detecting these neutrinos from NGC 1068 will give astronomers insight into the forces involved there, as well as into which specific engines are responsible for them–quite a boon given the hidden nature of black holes.

Fewer than 100 NGC 1068 neutrinos were detected at Earth, but they would have been diluted as they traveled across the vast volume of space. Accounting for this reduction, the astronomers say the total number of neutrinos generated by the black hole must be so huge that they carry away *a billion times* as much energy as the sun emits.

These observations also provide a major clue for another mystery. Neutrinos come to Earth from all over the sky, creating a background glow across the heavens. The source of this glow has been difficult to pin down. Neutrinos from several other active galaxies were also seen in the IceCube data (though with less statistical certainty than for NGC 1068), and there are many millions of these galaxies throughout the universe. The new data indicate that if they emit neutrinos much as NGC 1068 does, these more distant galaxies could be the source of the cosmic neutrino background, similar to how individual stars in the sky blur together to form the continuous glow of the Milky Way you can see from a dark site at night.

Not too long ago we knew of only two astronomical neutrino sources: the sun, where neutrinos are created in the nuclear fires of its core, and Supernova 1987A, a relatively nearby exploding star that emitted a transient flash of neutrinos once and then was gone.

Every big galaxy in the universe has a supermassive black hole in its core, and any of them can potentially be active. Yet, though ubiquitous, they can be difficult to observe. With a positive detection of neutrinos coming from at least one and probably several of them, astronomers have opened up a new window on these prodigious monsters.

This is an opinion and analysis article, and the views expressed by the author or authors are not necessarily those of Scientific American.

About the Author

Phil Plait is a professional astronomer and science communicator in Colorado. He writes the Bad Astronomy Newsletter.

Section 3: Dark Matter

Could Gravity's Quantum Origins Explain Dark Energy?

By Conor Purcell

For decades cosmologists have wondered about the nature of dark energy, the proposed antigravitational force behind the accelerating expansion of the universe. Since the 1990s astronomers have observed that the universe is not only expanding, but also increasing its expansion rate. This is very strange, because the collective gravitational pull of all the "stuff" in the universe would be expected to eventually reverse cosmic expansion, or at least slow it down. Instead, just like a ball gently tossed overhead suddenly soaring off into the heavens, some mysterious force–the aforementioned "dark energy"–is pushing far-distant, galaxy-filled regions of space away from us at ever-greater speeds. No known physics has fully explained this phenomenon; it remains a cosmic enigma, and its true, as-yet-unknown nature will profoundly shape the ultimate fate of our universe.

Now, however, a new theoretical study, submitted for publication at the *Journal for Cosmology and Astroparticle Physics*, suggests dark energy's apparent antigravitational properties may be the natural, inevitable consequence of how gravity works in the first place, at the universe's most fundamental quantum scales. If eventually verified by further cosmological evidence, the idea would represent a major breakthrough in the long quest to mend the schism between physicists' two most cherished theories: quantum mechanics, which describes the microscopic world of particles and fields, and general relativity, which describes the macroscopic cosmos of planets, stars and galaxies. General relativity posits that gravity is an emergent property of curves and warps in spacetime–the fabric of reality itself–but the theory loses its predictive power at quantum scales; conversely, quantum mechanics accurately incorporates all other known fundamental forces save for gravity, which fails to fit into the theory. Thus, many physicists suspect a quantum theory of gravity is the only way to unify these two opposing approaches.

According to Daniele Oriti, a co-author of the new paper, the core idea behind any theory of quantum gravity is that gravitation arises from a myriad of tiny, discrete, quantum objects that form a sort of hidden underworld, a deeper substructure beneath the familiar dimensions of space and time. "These quantum objects, which are very difficult to imagine," Oriti says, "are essentially the building blocks of space itself. They do not exist in space, but are themselves the very stuff out of which space is made. If they exist at all, they are absolutely tiny in their size, and are at a microscopic scale which even the most powerful microscopes cannot see."

In the study Oriti and his co-author Xiankai Pang, both at the University of Munich in Germany, focused first on developing a new quantum gravity model by trying to better understand the force's properties at the microscopic level. "Once having constructed our new model," Oriti says, "we decided to track it through time from the beginning of our modeled universe, to see what would happen during the evolution of its expansion. We were definitely surprised when we saw something closely resembling dark energy. The model produced an acceleration of the expansion of the universe at the stage corresponding to the time we are at today, which matches very closely with current observational evidence."

"This is quite an elegant result," says Abhay Ashtekar, an eminent theorist at Penn State who works on modern theories of quantum gravity and who was not involved in the new study. "Because the new approach begins with a general framework for quantum gravity at the subspace level, and then applies it to the cosmological scale, while in other methods one restricts oneself to the cosmological context right from start, the new idea is beginning from a more fundamental perspective than we have done before, and that is an advantage."

Oriti explains that the model's acceleration of the expansion of the universe, during the stage corresponding to today, is caused by interactions between the subspace quantum objects that make up gravity in the theory. After the expanding universe reaches a critical volume, these quantum objects begin to interact with each other in new ways. It is a bit like baking a cake. Imagine a cake where the yeast–in this case the subspace quantum objects–is not so important until a critical temperature–in this case the volume of the universe–is reached, whereafter conditions are just right to kick it into action, causing a rapid expansion. In the quantum gravity model, this is what causes the emergence of the dark energy–like phenomenon, which is characterized by an acceleration of the growth in volume of space.

"In the model, during the early universe, when the volume is small, the quantum objects out of which space emerges, interact in a manner which makes them subdominant compared to their large-scale long-term evolution," says Oriti. "But then, because the universe keeps expanding through time, at some point these interactions become relevant and they start affecting the evolution of the universe–the dynamics of the universe–in a considerable manner, causing an acceleration of the expansion. So, at that stage, the interactions between the quantum objects which make up space produce an acceleration which is similar in description and magnitude to the dark energy cosmologists observe."

"Having a dark energy phenomenological effect like this from a quantum gravity model is very interesting," says Ana Alonso Serrano, a physicist at the Max Planck Institute for Gravitational Physics in Germany, who was also not involved in the study. "I think it is important that we explore our quantum gravity models in this kind of way, so to see if they can make predictions about cosmology and compare them to observations."

"The next step will be to build on their theory, and their model, so to make further predictions which can be compared against real cosmological observations," she says. "But I think there is still a long road ahead before we really establish a good understanding about the quantum nature of gravity, and indeed if there is a firm relationship with dark energy."

About the Author

Conor Purcell is a science journalist who writes on science and its role in society and culture. He has a Ph.D. in Earth science and was the 2019 Journalist in Residence at the Max Planck Institute for Gravitational Physics in Germany.

Astronomers Might See Dark Matter by Staring into the Void

By Anil Ananthaswamy

In our search for cosmic signals of dark matter, we could be likened to drunkards looking for lost keys beneath lampposts, where the light shines the brightest. Here, the "lampposts" are regions of space teeming with galaxies and galaxy clusters, which are thought to be embedded in dense clouds, or "halos," of dark matter. What if instead we trained our sights on cosmic voids—vast reaches of mostly empty space? In a new preprint study, a trio of researchers argue that while dark matter's overall signal from these parts of the cosmos would be weaker, it would also be less contaminated by astrophysical sources and thus could be easier to spot.

"It's a fresh idea," says Nico Hamaus, a cosmologist at the Ludwig Maximilian University of Munich in Germany, who was not part of the study. "And it's not just the idea. It's also backed up with some calculations [that] make sense."

Dark matter is thought to make up more than 80 percent of the material in the universe. This estimate is based mostly on the gravitational influence this mysterious substance seems to exert on the gas, dust, stars and galaxies that make up normal matter. For instance, the rotational speeds of galaxies are such that, without the gravity of dark matter to hold them together, they would have long since disintegrated.

Physicists' best collective guess is that dark matter is made of so-called weakly interacting massive particles (WIMPs). But direct evidence for WIMPs has been elusive, despite decades of looking for them in particle accelerators and exquisitely sensitive detectors buried deep underground to minimize spurious signals from cosmic rays and other sources. Still, WIMPs remain the favored candidate for dark matter, says study co-author Nicolao Fornengo of the University of Torino in Italy.

According to almost all WIMP-based models, if these particles are, as expected, heavy–say, between a few giga-electron-volts (GeV) to a few tera-electron-volts (TeV), where a GeV is about the mass of a proton–then they should eventually decay or collide with one another and annihilate, which would produce gamma rays in both cases. "If dark matter produces [gamma rays], the signal should be in there," Fornengo says.

Current gamma-ray observatories, especially NASA's Fermi mission with its large area telescope (LAT), detect a diffuse all-sky "background" of gamma rays. This background is the unexplained excess that remains once the contributions from all known astrophysical sources, such as pulsars and matter-slurping supermassive black holes, are subtracted. And it isn't evenly distributed across the sky–which is consistent with what astrophysicists expect from dark matter emission and the astrophysical sources that are too small to be resolved with even the best-in-class Fermi LAT. When it comes to dark matter, the gamma-ray glow of decaying and annihilating WIMPs should correlate with large-scale cosmic structure, shining brighter from matter-packed regions and fainter from the voids. Early studies indicate that this correlation exists, but so far such studies have mostly avoided the voids and have focused instead on the brighter regions full of galaxies and clusters.

To see whether such a signal can be teased out better from voids than overdense regions, the team modeled how it should emanate from both types of cosmic structures. Their results suggest that, although the combined gamma-ray emission from dark matter and normal matter within a void would be much weaker than that from an overdense region, this weakness actually confers an advantage: the relative lack of normal matter ensures fewer astrophysical sources that would otherwise obscure dark matter's gamma-ray emission. "It's a trade-off between having a stronger but more polluted signal to measure versus a weaker but cleaner signal," Fornengo says. His and his colleagues' study has been submitted to the *Journal of Cosmology and Astroparticle Physics.*

The team also found—somewhat unsurprisingly—that most gamma rays from dark matter in these voids should emerge via the decay of the particles rather than their annihilation. In order for two particles to annihilate, they must first collide, and the odds of WIMPs finding each other in cosmic voids are low. But the particles should decay regardless of the density of their distribution. "Decay just probes the whole mass within [a volume of] space," Fornengo says. "And the mass of a void is not a small number. It's still a big object. It's just less dense."

Because of its superior signal-to-noise ratio and its bias toward detecting gamma rays from decaying particles, Hamaus says, the technique might offer novel insights into dark matter's properties that would be inaccessible via gamma-ray studies of overdense regions alone. For instance, the greater the average lifetime of a dark-matter particle, the less decay should occur over a given region of space and time. While such a faint signal would normally be undetectable, that should not be the case in voids. "Because your signal-to-background [noise] is higher, you can go further in the exploration of the parameter space," he says.

Anthony Pullen, an astrophysicist at New York University, who is unaffiliated with the study, is cautiously optimistic about near-future tests of its core ideas. Multiple large-scale surveys of cosmic structure are slated to commence later this decade on next-generation facilities such as the European Space Agency's (ESA's) Euclid space telescope, NASA's Nancy Grace Roman Space Telescope and the ground-based Vera C. Rubin Observatory. "As those surveys come online, you're going to have these very large data sets. The more galaxies we are able to detect, the better we'll be able to actually map out where the voids are," Pullen says. "And that would help with this type of study. In the next few years, you could see something like this demonstrated as a proof of concept."

Today such a proof of concept would have to rely on the gamma-ray data collected by Fermi LAT—which isn't up to the task, according to Fornengo and his colleagues. Making unequivocal detections, they calculate, would require a new generation of gamma-ray instruments with twice the detector volume and five times the angular resolution (the ability to tell apart sources in the sky) of Fermi LAT. "It would be a great addition to have a 'new Fermi,'" Fornengo says, although he acknowledges that, for now, such a detector only exists in their dreams. That hasn't stopped the team from giving it an appropriately Italian nickname, however: *Fermissimo*.

About the Author

Anil Ananthaswamy is author of The Edge of Physics *(Houghton Mifflin Harcourt, 2010),* The Man Who Wasn't There *(Dutton, 2015), and* Through Two Doors at Once: The Elegant Experiment That Captures the Enigma of Our Quantum Reality *(Dutton, 2018).*

Dark Matter Hunters Need Fresh Answers

By Joseph Howlett

Decades ago, astrophysicists brimmed with hope of discovering dark matter, the unseen mass that lets galaxies spin far faster than their stars' gravity would allow by itself. But underground traps set on Earth to capture supposed dark particles have now spent years measuring nothing but subterranean silence.

There should be five times as much dark matter around us as normal everyday matter, the stuff of stars, planets and people. So the failure to find this missing chunk of the universe–seen most recently in new results from the XENONnT experiment (in which I participated)–has sent physicists scrambling for an explanation. But most such attempts simply tweak an existing theory or turn to battle-worn alternatives just to propose more of the same: huge experimental efforts to build the most sensitive instruments ever, only for them to look at nothing.

If astrophysicists don't take big, new and different swings at dark matter, the search will only stagnate until it atrophies. Fortunately, the alarming reality that we're on the wrong track has led some back the drawing board to seek out new avenues for discovery rather than retreading old ground. These more ambitious, sometimes radical and farfetched, ideas, whether they succeed or not, at least attempt to reckon with the dire situation.

A Fuzzier Possibility

As Earth-bound detectors fail, desperate physicists have begun looking to the sky for answers. One new kind of dark matter search would, for example, look for galaxies with distortions symptomatic of "fuzzy dark matter." This theory considers dark matter particles as so incredibly lightweight that they act like waves instead of particles.

Rather than a tiny, localized bundle of mass shooting through the Earth like an invisible bullet, a fuzzy dark matter particle is everywhere in the galaxy at once, like how you can tune into the same satellite radio station in both New York and L.A.

On a galactic scale, fuzzy dark matter forms an underlying bath of undulating waves, buzzing together like a quiet static fuzz. Their combined gravity still holds the galaxy together, but the diffuse, fuzzy particles never smash into an atom like traditional particles, so you can never see them in a detector on Earth—which would explain our experimental failures thus far.

Versions of this idea have floated around for decades, but new energy has poured in recently as theorists realized that this kind of dark matter could have characteristic effects on galaxies that we can look for with the right kind of telescope. For instance, this wavy fuzz would form natural peaks and valleys across the galaxy as a result of the interference of so many wave-particles piled together. By carefully tracking the motion of stars in galaxies, we may be able to see signs of these variations, proving the rippling nature of the galactic glue.

These astronomical signatures make fuzzy dark matter especially attractive in an era of dwindling hope for underground dark matter searches.

Gravitational Detectors

Some ambitious experimentalists are starting to explore the use of modern quantum mechanical sensors to register the tiny gravitational pull of individual dark matter particles zooming through Earth. Dark matter's gravity, they fear, could be its only connection to the world of regular matter.

But we can only measure particles in a lab if they have another, much stronger interaction than gravity. Since gravity exerts a far gentler tug than any other force, the gravity of individual particles is too tiny to notice. You can only measure it in aggregate, like when all the particles of dark matter in a galaxy pull on a star in unison.

This makes gravity-only dark matter a kind of nightmare scenario, as it would be impossible to detect directly. Or so we thought.

Now scientists are trying to think of new technologies that would be sensitive to a much tinier gravitational pull than conventional detectors can see. Current experimental quantum mechanical sensors may only work if dark matter particles happen to be exceptionally heavy. But in the face of the scary but growing possibility that gravity-only dark matter is real, exploring this avenue of research is a necessity that might birth new, inventive technologies that can detect even lighter particles.

Back to Basics

A small but stalwart faction in physics has long held that it's gravity we don't understand, not the makeup of the universe. This idea is gaining new traction as dark matter particles evade detection and our main evidence remains gravitational.

One fresh idea in this arena comes from the Dutch theoretical physicist Erik Verlinde, who challenges the idea that gravity is a force at all. Instead, he starts with the observation that the second law of thermodynamics demands that the "entropy" of the universe, a number quantifying how "mixed up" things are, tends to increase. Verlinde claims that the phenomenon we call "gravity" is merely a result of this tendency—an apple on the ground somehow constitutes a more stirred-together world than an apple in a tree.

To explain this, he turns to a gravitational wonder: the black hole. We've known since the 1970s that gravity and entropy are related, thanks to black holes. As objects fall into a black hole and it grows larger, its entropy increases. But Verlinde really ran with this gravity-entropy connection, generalizing it to show that by cleverly rearranging the laws of entropy, you can get the familiar laws of gravity.

According to Verlinde, our mistaking of the laws of entropy for a separate force has led us to misunderstand the motion of stars around galaxies. His theory predicts all that movement without recourse to dark matter. The idea has been polarizing; critics have pointed out that the "entropic gravity" theory's predictions were incompatible with quantum measurements of very cold neutrons, only for other theorists to come to Verlinde's defense. But in an era of dwindling hope for dark matter, polarizing is good; more people should be thinking big and drawing heat like Verlinde.

Dark matter experiments of the past decade were built because of the beautiful simplicity of the hypothesis they hoped to prove. When they instead ruled it out, a cottage industry sprung up to invent more complicated versions of the hypothesis to explain its evasion of the current detectors, and justify their successors. The motivation is no longer simplicity, but inertia.

This is the nature of science; it's easier to pivot than completely shift gears. But at some point, it's time to go big or go home. Our best hope is to push in new directions, and even tear down long-standing pillars of theory, in order to resolve the most embarrassing hole in our picture of the universe.

This is an opinion and analysis article, and the views expressed by the author or authors are not necessarily those of Scientific American.

About the Author

Joseph Howlett is a postdoctoral physicist at Stanford University. While working toward his Ph.D. at Columbia University, he participated in a collaboration searching for dark matter with a large underground detector.

Stars Made of Antimatter Might Be Lurking in the Universe

By Leto Sapunar

Antimatter may seem like the stuff of science fiction—especially because scarcely any of it can be seen in our universe, despite physicists' best theories suggesting antimatter should have arisen in equal proportion to normal matter during the big bang. But researchers do regularly produce particles of antimatter in their experiments, and they have the inklings of an explanation for its cosmic absence: Whenever antimatter and normal matter meet, they mutually annihilate in a burst of energy. The slimmest overabundance of normal matter at the beginning of time would have therefore effectively wiped antimatter off the celestial map, save for its occasional production in cosmic-ray strikes, human-made particle accelerators and perhaps certain theorized interactions between particles of dark matter.

That is why physicists were so greatly puzzled back in 2018, when the head of the Alpha Magnetic Spectrometer (AMS) experiment mounted on the exterior of the International Space Station announced that the instrument might have detected two antihelium nuclei—in addition to six that were possibly detected earlier. Any way you slice it, known natural processes would struggle to produce enough antihelium for any of it to end up in our space-based detectors. But the easiest of all those hard methods would be to cook up the antihelium inside antistars—which, of course, do not seem to exist. Despite the fact that the entirely unexpected AMS results have yet to be confirmed, let alone formally published, scientists have taken them seriously, and some have scrambled to find explanations.

Inspired by the tentative AMS findings, a group of researchers recently published a study calculating the maximum number of antimatter stars that could be lurking in our universe, based on a count of currently unexplained gamma-ray sources found by the Fermi Large Area Telescope (LAT). Simon Dupourqué, the study's lead author and an astrophysics graduate student at the Research Institute in Astrophysics and Planetology at the University of Toulouse III–Paul Sabatier in France and the French National Center for Scientific Research (CNRS), made the estimate after looking for antistar candidates in a decade's worth of the LAT's data.

Antistars would shine much as normal ones do—producing light of the same wavelengths. But they would exist in a matter-dominated universe. As particles and gases made of regular matter fell into such a star's gravitational pull and made contact with its antimatter, the resulting annihilation would produce a flash of high-energy light. We can see this light as a specific color of gamma rays. The team took 10 years of data, which amounted to roughly 6,000 light-emitting objects. They pared the list down to sources that shone with the right gamma frequency and that were not ascribed to previously cataloged astronomical objects. "So this left us with 14 candidates, which, in my opinion and my co-authors' opinion, too, are *not* antistars," Dupourqué says. If all of those sources were such stars, however, the group estimated that about one antistar would exist for every 400,000 ordinary ones in our stellar neck of the woods.

In place of any putative antistars, Dupourqué says, these gamma flashes could instead be coming from pulsars or the supermassive black holes at the centers of galaxies. Or they might simply be some kind of detector noise. The next step would be to point telescopes at the locations of the 14 candidate sources to find out if they resemble a star or a prosaic gamma-emitting object.

Given some interesting but questionable gamma sources, calculating the conceivable "upper limit" to the number of antistars is a long shot from actually discovering such astrophysical objects, So most researchers are not leaning toward that conclusion. "According to both theory and observations of extragalactic gamma rays, there should be no antistars in our galaxy.... One would only expect upper limits consistent with zero," says Floyd Stecker, an astrophysicist at NASA's Goddard Space Flight Center, who was not involved in the research. "However, it is always good to have further observational data confirming this."

If scientists, including the authors, are skeptical of antistars' very existence, why are they worth discussing? The mystery lies in those pesky possible detections of antihelium made by the AMS, which remain unexplained. Antiparticles can be created from two known natural sources—cosmic rays and dark matter—but the odds that either of them are responsible appear to be vanishingly slim.

As we increase the size of an atom, it becomes harder and harder to produce as an antiparticle, says Vivian Poulin, a CNRS cosmologist based in Montpelier, France. This "means that it's rarer and rarer that it occurs, but it's allowed by physics." An antiproton is relatively easy to form, yet anything heavier, such as antideuterium—an antiproton plus an antineutron—or antihelium—two antiprotons plus typically one or two antineutrons—gets progressively harder to make as it gets more massive. In a paper published in 2019, Poulin used the AMS's potential antihelium detections to calculate a rough estimate of the prevalence of antistars, which inspired Dupourqué's new study.

In a process called spallation, high-energy cosmic rays from exploding stars can ram into interstellar gas particles, says Pierre Salati, a particle astrophysicist at the Annecy-le-Vieux Particle Physics Laboratory, who worked on Poulin's 2019 study. The team responsible for the AMS's antiparticle detections claim it may have detected six antihelium-3 nuclei, which would be incredibly rare products of spallation, and two antihelium-4 nuclei, which would be almost statistically impossible to form from cosmic rays, Salati says. (The difference between the two isotopes is the addition of one antineutron.)

As for dark matter, certain models predict that dark matter particles can annihilate one another—a process that could also create antiparticles. But this process still might not be able to make antihelium-4 in high enough quantities for us to have a realistic chance of ever seeing it (if such speculative models reflect reality at all). That is why the antistar hypothesis is still on the table.

Verified antihelium detections would be a good indicator for the existence of antistars, but so far the AMS is the lone experiment to offer any such evidence–which has yet to be granted peer-reviewed publication, Salati notes.

"It's a very challenging analysis because, for every one antihelium event, there are 100 million regular helium events," says Ilias Cholis, an astrophysicist at Oakland University, who also worked on Poulin's study. It is possible, he and others say, that the detections turn out to be a fluke of a very complicated analysis.

Samuel Ting, a Nobel laureate physicist at the Massachusetts Institute of Technology, heads the AMS team and first publicly presented the two latest possible antihelium detections–the antihelium-4 candidates–in 2018. "We are not yet ready to publish any heavy antimatter results," he says. "We are collecting more data before any [further] announcement is made."

It is possible that a different experiment may give answers sooner. The General AntiParticle Spectrometer (GAPS) experiment is a balloon-borne detector that will hunt for antiparticles above Antarctica this year. Finding more antiparticles–antideuterons or even antihelium, in particular, according to Cholis–with the GAPS detector would make the AMS results far more convincing.

If antistars were found to be the culprit, that discovery would require a major reenvisioning of the universe's evolution: no longer could we relegate antistars and other hypothetical astrophysical objects composed of antimatter to the fringes of reasonable speculation. Even if they do exist, however, antistars probably are not forming now, Salati says, because their presumptive natal clouds of antihydrogen would face steep odds of avoiding annihilation for the past 13 billion years or so. Thus, any antistars that might be found likely would be exceedingly old remnants of the early universe. If so, one deep mystery would be replaced with another: How, exactly, did such ancient relics manage to survive to today? As is often the case, a new discovery raises far more questions than it answers.

About the Author

Leto Sapunar is a freelance science journalist from the Pacific Northwest. You can also find his bylines at Popular Science *and* Retraction Watch.

Are Telescopes the Only Way to Find Dark Matter?

By Chanda Prescod-Weinstein

"How do you think the dark matter problem is solved?" Vera C. Rubin urgently asked me, within minutes of being introduced at a 2009 Women in Astronomy conference. To this day, I can't remember what I said in response. I was awestruck: the famed astronomer who had won the National Medal of Science for her work finding the first conclusive evidence for dark matter's existence was asking *me*, a twentysomething Ph.D. student, for my opinion. I am certain that whatever I came up with was not very good because it was a problem that I had, until that moment, given no serious thought to. Until Rubin asked me my opinion, it had never occurred to me that I was entitled to have an opinion on the question at all.

If I disappointed her with my answer, she didn't show it. Instead she asked me to sit down to lunch with her and some other women astronomers, including former NASA administrator Nancy Grace Roman. Rubin then proceeded to fangirl over Roman, who is often referred to as "the mother of the Hubble Space Telescope." It was quite a moment for me, to watch an elderly woman who had uncovered one of the greatest scientific mysteries of our time excitedly introduce us to her own hero.

Rubin cemented her legacy in the 1960s, when she studied stars inside galaxies and found something odd: stars on the outskirts of galaxies were moving faster than they were supposed to, as if there was an invisible matter there contributing a gravitational pull. Her work echoed findings from galaxy cluster studies in the early 1930s by Fritz Zwicky, which had led him to suggest the existence of *Dunkle Materie*, German for "dark matter." Throughout the 1970s Rubin and astronomer Kent Ford published data consistent with this conclusion, and by the early 1980s scientists were in widespread agreement that physics had a dark matter problem.

Most attempts to track down dark matter in the laboratory have fallen into three categories. So-called direct detection experiments look for evidence of dark matter particles interacting with particles of normal matter—for instance, the element xenon—through one of the nongravitational fundamental forces, the weak force, as well as through hypothesized new forces. Collider experiments, such as those at the Large Hadron Collider near Geneva, take the opposite approach, smashing two regular particles together with the hope of producing dark matter particles. Meanwhile "indirect detection" experiments look for evidence of dark matter interacting with itself, with the resulting collision producing observable particles.

So far none of these strategies has turned up the missing matter. We still don't know if dark matter can talk to regular matter in any way beyond gravity. It may be impossible to produce in the accelerators we can build or to detect in the experiments we can construct. For this reason, astronomical observations—cosmic probes of dark matter—are one of our best hopes. These probes allow us to look for signatures of dark matter in environments that are difficult for us to produce on Earth—for example, inside neutron stars. More broadly, such searches look at dark matter's behavior under gravity in a variety of locations.

Despite the promise of this approach for studying dark matter, it has sometimes been caught in the middle between the astronomy and physics communities. Physicists tend to emphasize colliders and laboratory experiments and don't always prioritize links to astrophysical work. Astronomers tend to write dark matter off as a particle physics problem. This disconnect has implications for funding. In 2022 we have an opportunity to change that. The start of the 2020s marked the beginning of an important process known as the Snowmass Particle Physics Community Planning Exercise. This project, which takes place about once a decade, brings physicists together to explain prospective scientific projects to a congressionally mandated panel that will determine scientific priorities. For the first time, cosmic probes of dark matter will be a distinct topic of consideration. Although Snowmass does not make formal policy recommendations, it is certainly the case that at each stage of the organizational hierarchy, there will be decisions about what science to emphasize.

Cosmic probes allow us to look for signatures of dark matter in environments that are difficult for us to produce on Earth– for example, inside neutron stars.

A Universe of Dark Matter Candidates

There is still much we don't know about dark matter, but we have come a long way since Rubin's work in the 1970s and 1980s. We now have good evidence to suggest that every galaxy lives in its own bubble of dark matter–called a dark matter halo–that extends well beyond the visible part of the galaxy. The amount of dark matter in these galaxy-halo systems surpasses the amount of matter in the stars and planets and gas. In other words, all the particles that we have been able to identify in labs and colliders–referred to collectively as the Standard Model of particle physics–contribute only about 20 percent of the normally gravitating matter in the universe. If we take into account dark energy and the fact that matter and energy are fundamentally equivalent, we are down to understanding only about 4 percent of the cosmos. The Standard Model is both a stunning achievement and a theory that is, apparently, deeply incomplete. We need a new particle or particles to solve the problem.

Physicists now have an assortment of dark matter candidates. Most scientists favor candidates that are "cold dark matter"–particles that move relatively slowly (meaning, at nonrelativistic speeds much slower than that of light). Within the class of cold dark matter, one of the classic models is the weakly interacting massive particle (WIMP). Scientists presume that WIMPs would have formed naturally in the early universe and predict that they have some kind of interaction with regular matter through the weak force. The most popular WIMP models fall into a category of particles called fermions–a class that includes electrons and quarks.

WIMPs were the most highly favored dark matter candidates for a long time, particularly in the U.S. Opinions have shifted in recent years, though, as evidence for WIMPs has failed to show up at the Large Hadron Collider or in any of the direct and indirect detection experiments.

Recently the particle physics community has become excited about another hypothetical dark matter candidate: an axion. Axions are predicted to have smaller masses than WIMPs, and they are not fermions. Instead axions belong to a class of particles called bosons–the category that includes photons, or particles of light. As bosons, axions have fundamentally different properties than WIMPs, which opens the door to an intriguing possibility about the structures they could form. Axions are what first drew me into the world of dark matter research.

Alluring Alternatives to WIMPs

Five years passed between my conversation with Vera Rubin and my first attempt at answering the question she had put to me. By then it was 2014, and I was a Dr. Martin Luther King, Jr., postdoctoral fellow at the Massachusetts Institute of Technology, appointed first to the Kavli Institute for Astrophysics and Space Research and then the Center for Theoretical Physics (CTP) and looking for something interesting to work on. It was there that Mark Hertzberg–at the time also a postdoctoral researcher at CTP–and I first started talking about a debate that had erupted among physicists: Could axions form an exotic state known from atomic physics called a Bose-Einstein condensate?

This possibility arises from a fundamental difference between bosons and fermions. Fermions must obey the Pauli exclusion principle, which means two fermions cannot share the same quantum state. This rule is why electron orbitals in chemistry can be so complicated: because the electrons orbiting an atom cannot occupy the same quantum state, they must spread out in different patterns with different amounts of energy called orbitals. Axions, on the other hand, can share a quantum state. This means that when we cool them enough they can all enter the same low-energy state and act collectively like one superparticle—a Bose-Einstein condensate. The possibility that this could happen naturally in space is, in my view, quite exciting.

Axions had been proposed in the 1970s by Hertzberg's Ph.D. adviser at M.I.T., Frank Wilczek, one of the first to realize that one consequence of a model proposed by Helen Quinn and the late Roberto Peccei was a particle, which Wilczek named "axion" after a brand of laundry detergent. Thus, Hertzberg was already quite familiar with axions. I, on the other hand, was relatively new to this idea. I had spent most of my career focused on other questions, and I had to get up to speed. Along the way, I learned to distinguish between the traditional axion and the class of particles that physicists have come to loosely refer to as axionlike particles.

The traditional axion arises from the Peccei-Quinn extension to the theory of quantum chromodynamics (QCD), which describes another of the four fundamental forces, the strong force. Although QCD is a highly successful model, it also predicts phenomena we've never observed. Peccei and Quinn's work solves this problem, while providing a mechanism for producing dark matter. But another idea called string theory also proposes a series of particles with the same mathematical structure as the original axion; these particles have come to be called axionlike. The traditional QCD axion is usually expected to have a mass of about 10–35 kilogram—several orders of magnitude lighter than the electron—but the larger class of axions from string theory can be much lighter, down to 10–63 kilogram.

The work Hertzberg and I did together with our postdoctoral adviser Alan Guth led us to quibble with a popular view of how axions might form Bose-Einstein condensates. A distinguished physicist, Pierre Sikivie of the University of Florida, had prompted much excitement in 2009, when he proposed that QCD axions would form large condensates in the very early universe. His calculations suggested they would lead to ringlike galaxy halos rather than the spherical halos that most astronomers expect and that WIMP models predict. If so, then we might be able to tell what dark matter is made of just by looking at halo shapes.

But when Mark, Alan and I sat down to check how Sikivie's group had arrived at this prediction, we came to a radically different conclusion. Although we agreed that axion Bose-Einstein condensates would form in the early universe, they would be much smaller—the size of asteroids. Our model also did not give any indications, in the present-day universe, of what kind of axion structures we might find billions of years in the future. Trying to better model how—and whether—we get from small asteroid-sized condensates to the galactic-scale dark matter halos of today is still a significant computational challenge.

The same year our paper came out, another group was looking into other interesting implications of axionlike particles. A team led by Hsi-Yu Schive of National Taiwan University published computer simulations of certain axionlike particles that are often referred to as "ultralight axions" or "fuzzy dark matter," so named because they have a very low mass and would act like blurred-out waves rather than pointlike particles. They showed that these particles could form wavelike dark matter halos with Bose-Einstein condensates at their cores. Schive's paper generated new interest in ultralight axions and raised hopes that astrophysical observations could detect signs of the wavelike halo structures we expect.

Doing science is never just about calculations, observations and experiments; it is also about working collaboratively with other people, including policy makers.

These days axions and axionlike particles stand along with WIMPs as some of our best guesses at what dark matter could be. Another category that is growing in popularity is a model called self-interacting dark matter (SIDM). This idea predicts fermion dark matter particles that have some kind of interaction with one another—a self-interaction—beyond gravity. These self-interactions could create more interesting shapes and structures within a halo than a smooth, spherical blob. The particulars of the structures are hard to predict, though, and depend on the mass and other characteristics of the particles. Interestingly, axions may also interact with one another, though in different ways than self-interacting fermions.

There is an alternative to WIMPs, axions and SIDM: neutrinos. Although Standard Model neutrinos are now known to be too low in mass to explain all of the missing matter, these neutrinos are real and hard to see, making them functionally a small component of the dark matter that we call the cosmic neutrino background. In addition, a new type of neutrino has been hypothesized as a companion to the Standard Model neutrino: the sterile neutrino. Sterile neutrinos are distinct because they interact primarily gravitationally and only mildly through Standard Model forces. In addition, they are perhaps the most popular warm–or at least somewhere between hot and cold–dark matter proposal.

Another idea that theorists are just starting to explore is that rather than a single dark matter particle, there may be an entire sector. Perhaps dark matter is made of traditional axions, axionlike particles, WIMPs, sterile neutrinos and SIDM–all together. One other tantalizing possibility is that dark matter actually comprises stellar-mass black holes that would have formed in the early universe. This option has become more popular since the 2017 detection of gravitational waves indicated that black holes in this mass range are more common than expected.

Clues in the Sky

In astronomy we are relatively passive observers. We can choose our instruments, but we cannot design a galaxy or a stellar process and watch it unfold. Cosmic phenomena rarely happen on human-friendly time scales–galaxy formation takes billions of years, and the cosmic processes that might emit dark matter particles do so over tens to hundreds of years.

Even so, astrophysical probes of dark matter can tell us a lot. For instance, the NASA Fermi Gamma-ray Space Telescope has functioned as a dark matter experiment by looking for gamma-ray signatures that could be explained only by dark matter. WIMPs, for instance, are predicted to be their own antimatter partners, meaning that if two WIMPs collided, they would annihilate each other just as matter and antimatter do on contact. These explosions should produce an abundance of gamma-ray light where there is dark matter, especially at the cores of galaxies where dark matter is densest.

In fact, the Fermi telescope does see an excess of gamma-ray light at the center of the Milky Way. These observations have inspired passionate debate among observers and theorists. One interpretation is that these fireworks result from dark matter colliding with itself. Another possibility is that the signal comes from neutron stars near the center of the Milky Way that emit gamma-ray light through the typical course of their lives. Some astrophysicists favor the more mundane neutron star explanation, but others think the signal is dark matter. The fact that there is disagreement is normal, and even I have a hard time deciding what I think. I am compelled by physicists Tracy Slatyer and Rebecca Leane's thoughtful research showing that a dark matter explanation is sensible, but in the end, only analysis of more detailed observations will persuade the community about either idea. Future data from the Fermi telescope and proposed experiments such as NASA's All-sky Medium Energy Gamma-ray Observatory eXplorer (AMEGO-X for short) have the potential to settle the debate.

Scientists have also used the Fermi telescope to look for evidence of axions. Theories predict that when axions encounter magnetic fields, they occasionally decay into photons. We hope that by looking over long distances, we might see signs of this light, offering proof that axions exist.

And neutron stars–the potential confounding signal at the Milky Way's center–are actually a good place to look for dark matter on their own. Some theories suggest that these dense spinning stars produce axions when protons and neutrons collide in their cores. We might be able to observe these axions as they decay into photons and escape from the stars. And as neutron stars release the dark matter over tens to hundreds of years, they would cool down in a pattern that we may be able to measure–if we look long enough. Another hot topic of study right now is whether nonaxion dark matter collects in neutron stars, affecting the structure of the star. As a member of the NASA Neutron Star Interior Composition Explorer (NICER) collaboration, I am leading a research project that is using data from NICER, a little telescope on the International Space Station that is up for renewal later this year. Our project is looking for evidence that dark matter is inside or enveloping neutron stars.

We can also learn more about the nature of dark matter by studying the best evidence we have for its existence so far–the cosmic microwave background (CMB) radiation. This light is a radio signal that originated in the early universe, and it is inescapably everywhere, all around us. It provides a snapshot of a moment early in cosmic history, and the patterns we see in the frequencies of its light reflect the makeup of the universe when it was created. It turns out that we can only explain the patterns we see in the CMB by assuming that dark matter was present–if there were no dark matter, the CMB data would make no sense. The patterns in the data tell us what fraction of the total mass and energy dark matter contributed; they even help constrain the possible masses of the dark matter particles. As I write, the CMB-Stage 4 collaboration is preparing to use a collection of telescopes in Chile's Atacama Desert and at the South Pole to take the most detailed measurements yet of the CMB.

On the Horizon

Since that 2009 Women in Astronomy conference, Rubin and Roman have both passed away, but their legacies live on through projects that will seek to better understand our universe. The NASA Nancy Grace Roman Space Telescope will launch in the mid-2020s, and although it is primarily focused on studying cosmic acceleration (the "dark energy problem") and exoplanets, it will also provide insight into dark matter. At the same time, here on Earth, the Vera C. Rubin Observatory in the Atacama Desert will support research on many questions, including the search for the dark matter that made Rubin famous.

In other words, we have lots to look forward to in the coming years. One reason is that almost any large-scale astronomical observation has something to tell us about dark matter. For example, a team in Mexico led in part by Alma X. Gonzalez-Morales and Luis Arturo Ureña-López showed that we can use the phenomenon of gravitational lensing, where large masses bend spacetime so much that it acts like a fun-house mirror, to place constraints on the mass of fuzzy dark matter. Gonzalez-Morales and Ureña-López are both active participants in the Rubin Observatory's Legacy Survey of Space and Time program, working on gravitational lensing and participating in the dark matter working group. Within the group, we are discussing observations that will capture more detailed information about dark matter halos that can then be compared with computer simulations of proposed dark matter candidates. Similarly, Roman telescope surveys of large-scale structure will provide insight into dark matter's behavior on cosmic scales.

In the future, proposed x-ray observatories such as the NASA Spectroscopic Time-Resolving Observatory for Broadband Energy X-rays (STROBE-X) can help us take a closer look at neutron star structure in ways that will enhance our understanding of dark matter's possible properties. Other proposed future projects such as NASA's All-sky Medium Energy Gamma-ray Observatory, or AMEGO (not to be mistaken for AMEGO-X), will do the same in a different wavelength.

I will be an active participant not just as a scientist but as one of three conveners, alongside Alex Drlica-Wagner and Hai-Bo Yu, for the Snowmass Cosmic Frontier's topic Dark Matter: Cosmic Probes. It is our responsibility to describe the excitement and possibilities of astrophysical searches for dark matter to the funding decision-makers. The document I will help produce may influence guidance that is given to the National Science Foundation and the U.S. Department of Energy about what research we conduct over the next decade.

Coincidentally, the astronomy community just recently completed a similar process known as the 2020 Decadal Survey on Astronomy and Astrophysics. The resulting report sidestepped substantively addressing the dark matter problem, but it still offered strong support for efforts to better map the CMB, instruments to study neutron stars and x-ray observatories–three goals that will help us understand dark matter.

Doing science is never just about calculations, observations and experiments; it is also about working collaboratively with other people, including policy makers. How much progress we make will depend in part on what kind of support we get from lawmakers. Of course, this is stressful to think about. The good news is that there is a universe to wonder about, and trying to understand dark matter is a great distraction.

About the Author

Chanda Prescod-Weinstein is a theoretical physicist. She is an assistant professor of physics and a core faculty member in women's and gender studies at the University of New Hampshire. She is author of The Disordered Cosmos: A Journey into Dark Matter, Spacetime, and Dreams Deferred *(Bold Type Books, 2021).*

Section 4: Entanglement

String Theory May Create Far Fewer Universes Than Thought

By Clara Moskowitz, Lee Billings

The problem with string theory, according to some physicists, is that it makes too many universes. It predicts not one but some 10,500 versions of spacetime, each with its own laws of physics. But with so many universes on the table, how can the theory explain why ours has the features it does?

Now some theorists suggest most–if not all–of those universes are actually forbidden, at least if we want them to have stable dark energy, the supposed force accelerating the expansion of the cosmos. To some, eliminating so many possible universes is not a drawback but a major step forward for string theory, offering new hope of making testable predictions. But others say the multiverse is here to stay, and the proposed problem with all those universes is not a problem at all.

The debate was a hot topic in the summer of 2018 in Japan, where string theorists convened for the conference Strings 2018. "This is really something new and it's led to a controversy within the field," says Ulf Danielsson, a physicist at Uppsala University in Sweden. The conversation centers on a pair of papers posted on the preprint server arXiv in June 2018 taking aim at the so-called landscape of string theory–the incomprehensible number of potential universes that result from the many different solutions to string theory's equations that produce the ingredients of our own cosmos, including dark energy. But the vast majority of the solutions found so far are mathematically inconsistent, the papers contend, putting them not in the landscape but in the so-called swampland of universes that cannot actually exist. Scientists have known many solutions must fall in this swampland for years, but the idea that most, or maybe all, of the landscape solutions might live there would be a major change. In fact, it may be theoretically impossible to find a valid solution to string theory that includes stable dark energy, says Cumrun Vafa, a Harvard University physicist who led the work on the two papers.

Lost in the Multiverse

String theory is an attempt to describe the whole universe under a single "theory of everything" by adding extra dimensions of spacetime and thinking of particles as minuscule vibrating loops. Many string theorists contend it is still the most promising direction for pursuing Albert Einstein's dream of uniting his general theory of relativity with the conflicting microscopic world of quantum mechanics. Yet the notion of a string theory landscape that predicts not just one universe but many has put some physicists off. "If it's really the landscape, in my view it's death for the theory because it loses all predictive value," says Princeton University physicist Paul Steinhardt, who collaborated on one of the recent papers. "Literally anything is possible." To Steinhardt and others, the newfound problems with dark energy offer string theory a way out. "This picture with a big multiverse could be mathematically wrong," Danielsson says. "Paradoxically this makes things much more interesting because that means string theory is much more predictive than we thought it was."

Some string theorists such as Savdeep Sethi of the University of Chicago welcome the reevaluation that is happening now. "I think this is exciting," he says. "I've been a skeptic of the landscape for a long time. I'm really happy to see the paradigm shift away from this belief that we have this proven set of solutions." But not everyone buys the argument that the landscape actually belongs in the swampland—especially the research team that established one of the earliest versions of the landscape in the first place back in 2003, which goes by the acronym KKLT after the scientists' last names. "I think it's very healthy to make these conjectures and check what other things could be going on, but I don't see either theoretical or experimental reasons to take such a conjecture very seriously," says KKLT member Shamit Kachru of Stanford University. And Eva Silverstein, a Stanford physicist who also helped build the early landscape models, likewise doubts Vafa and his colleagues' argument. "I think the ingredients KKLT use and the way they put them together is perfectly valid," she says. Juan Maldacena, a theorist at the Institute for Advanced Study, says he also still supports the idea of string theory universes with stable dark energy.

And many theorists are perfectly happy with the string theory multiverse. "It is true that if this landscape picture is correct, the bit of the universe we're in compared to the multiverse will be like our solar system within the universe," Kachru says. And that is a good thing, he adds. Johannes Kepler originally sought a fundamental reason for why Earth lies the distance it does from the sun. But now we know the sun is just one of billions of stars in the galaxy, each with its own planets, and the Earth-sun distance is simply a random number rather than a result of some deep mathematical principle. Likewise, if the universe is one of trillions within the multiverse, the particular parameters of our cosmos are similarly random. The fact that these numbers seem perfectly fine-tuned to create a habitable universe is a selection effect–humans will of course find themselves in one of the rare corners of the multiverse where it is possible for them to have evolved.

The Accelerating Universe

If it is true that string theory cannot accommodate stable dark energy, that may be a reason to doubt string theory. But to Vafa, it is a reason to doubt dark energy–that is, dark energy in its most popular form, called a cosmological constant. The idea originated in 1917 with Einstein and was revived in 1998 when astronomers discovered that not only is spacetime expanding–the rate of that expansion is picking up. The cosmological constant would be a form of energy in the vacuum of space that never changes and counteracts the inward pull of gravity. But it is not the only possible explanation for the accelerating universe. An alternative is "quintessence," a field pervading spacetime that can evolve. "Regardless of whether one can realize a stable dark energy in string theory or not, it turns out that the idea of having dark energy changing over time is actually more natural in string theory," Vafa says. "If this is the case, then one can measure this sliding of dark energy by astrophysical observations currently taking place."

So far all astrophysical evidence supports the cosmological constant idea, but there is some wiggle room in the measurements. Upcoming experiments such as Europe's Euclid space telescope, NASA's Wide-Field Infrared Survey Telescope (WFIRST) and Chile's Simons Observatory being built in the desert will look for signs that dark energy was stronger or weaker in the past than the present. "The interesting thing is that we're already at a sensitivity level to begin to put pressure on [the cosmological constant theory]," Steinhardt says. "We don't have to wait for new technology to be in the game. We're in the game now." And even skeptics of Vafa's proposal support the idea of considering alternatives to the cosmological constant. "I actually agree that [a changing dark energy field] is a simplifying method for constructing accelerated expansion," Silverstein says. "But I don't think there's any justification for making observational predictions about the dark energy at this point."

Quintessence is not the only other option. In the wake of Vafa's papers, Danielsson and his colleagues proposed another way of fitting dark energy into string theory. In their vision, our universe is the three-dimensional surface of a bubble expanding within a larger-dimensional space. "The physics within this surface can mimic the physics of a cosmological constant," Danielsson says. "This is a different way of realizing dark energy compared to what we've been thinking so far."

A Beautiful Theory

Ultimately the debate going on in string theory centers on a deep question: What is the point of physics? Should a good theory be able to explain the particular characteristics of the universe around us or is that asking too much? And when a theory conflicts with the way we think our universe works, do we abandon the theory or the things we think we know?

String theory is incredibly appealing to many scientists because it is "beautiful"–its equations are satisfying and its proposed explanations elegant. But so far it lacks any experimental evidence supporting it—and even worse, any reasonable prospects for gathering such evidence. Yet even the suggestion that string theory may not be able to accommodate the kind of dark energy we see in the cosmos around us does not dissuade some. "String theory is so rich and beautiful and so correct in almost all the things that it's taught us that it's hard to believe that the mistake is in string theory and not in us," Sethi says. But perhaps chasing after beauty is not a good way to find the right theory of the universe. "Mathematics is full of amazing and beautiful things, and most of them do not describe the world," physicist Sabine Hossenfelder of the Frankfurt Institute for Advanced Studies wrote in her recent book, *Lost in Math: How Beauty Leads Physics Astray* (Basic Books, 2018).

Despite the divergence of opinions, physicists are a friendly bunch and are united by their common goal of understanding the universe. Kachru, one of the founders of the landscape idea, worked with Vafa, the landscape's critic, as his undergraduate adviser—and the two are still friends. "He asked me once if I'd bet my life these [landscape solutions] exist," Kachru says. "My answer was, 'I wouldn't bet my life, but I'd bet his!'"

Referenced

On the Cosmological Implications of the String Swampland. Prateek Agrawal, Georges Obied, Paul J. Steinhard and Cumrun Vafa in *Physics Letters B*, Vol. 784, pages 271–276; September 10, 2018.

Emergent de Sitter Cosmology from Decaying Anti–de Sitter Space. Souvik Banerjee, Ulf Danielsson, Giuseppe Dibitetto, Suvendu Giri and Marjorie Schillo in *Physical Review Letters*, Vol. 121, Article 261301; December 27, 2018.

De Sitter Space and the Swampland. Georges Obied, Hirosi Ooguri, Lev Spodyneiko and Cumrun Vafa (in press). Preprint available at arXiv:1806.08362v3

About the Author

Clara Moskowitz is a senior editor at Scientific American, *where she covers space and physics.*

Lee Billings is a senior editor for space and physics at Scientific American.

Tangled Up in Spacetime

By Clara Moskowitz

"All the world's a stage...," Shakespeare wrote, and physicists tend to think that way, too. Space seems like a backdrop to the action of forces and fields that inhabit it but space itself is not made of anything—or is it? Lately scientists have begun to question this conventional thinking and speculate that space—and its extension according to general relativity, spacetime—is actually composed of tiny chunks of information. These chunks might interact to create spacetime and give rise to its properties, such as the concept that curvature in spacetime causes gravity. If so, the idea might not just explain spacetime but might help physicists achieve a long-sought goal: a quantum theory of gravity that can merge general relativity and quantum mechanics, the two grand theories of the universe that tend not to get along. Lately the excitement of this possibility has engrossed hundreds of physicists who have been meeting every three months or so under the banner of a project dubbed "It from Qubit."

The "it" in this case is spacetime, and the qubit (pronounced "cue-bit," from "quantum bit") represents the smallest possible amount of information—a computer "bit" on a quantum scale. The idea suggests the universe is built up from some underlying code, and that by cracking this code, physicists will finally have a way to understand the quantum nature of large-scale events in the cosmos. The most recent It from Qubit (IfQ) meeting was held in July at the Perimeter Institute for Theoretical Physics in Ontario, where organizers were expecting about 90 registrants. Instead, they got so many applications they had to expand to take 200 and simultaneously run five satellite sessions at other universities where scientists could participate remotely. "I think this is one of the most, if not the most, promising avenues of research toward pursuing quantum gravity," says Netta Engelhardt, a postdoctoral researcher at Princeton University who is not officially involved in It from Qubit but who has attended some of its meetings. "It's just taking off."

Because the project involves both the science of quantum computers and the study of spacetime and general relativity, it brings together two groups of researchers who do not usually tend to collaborate: quantum information scientists on one hand and high-energy physicists and string theorists on the other. "It marries together two traditionally different fields: how information is stored in quantum things and how information is stored in space and time," says Vijay Balasubramanian, a physicist at the University of Pennsylvania who is an IfQ principal investigator. About a year ago the Simons Foundation, a private organization that supports science and mathematics research, awarded a grant to found the It from Qubit collaboration and finance physicists to study and hold meetings on the subject. Since then excitement has grown and successive meetings have drawn in more and more researchers, some official members of the collaboration funded by Simons and many others simply interested in the topic. "This project is addressing very important questions, but very difficult questions," says IfQ collaborator Beni Yoshida, a postdoctoral researcher at Perimeter. "Collaboration is necessary–it's not like a single person can solve this problem." Even scientists outside of the project have taken notice. "If the link with quantum information theory proves as successful as some anticipate, it could very well spark the next revolution in our understanding of space and time," says string theorist Brian Greene of Columbia University, who is not involved in IfQ. "That's a big deal and hugely exciting."

Entangling Spacetime

The notion that spacetime has bits or is "made up" of anything is a departure from the traditional picture according to general relativity. According to the new view, spacetime, rather than being fundamental, might "emerge" via the interactions of such bits. What, exactly, are these bits made of and what kind of information do they contain? Scientists do not know. Yet intriguingly, "what matters are the relationships" between the bits more than the bits themselves, says IfQ collaborator Brian Swingle, a postdoc at Stanford University. "These collective relationships are the source of the richness. Here the crucial thing is not the constituents but the way they organize together."

The key to this organization may be the strange phenomenon known as quantum entanglement–a weird kind of correlation that can exist between particles, wherein actions performed on one particle can affect the other even when a great distance separates them. "Lately one absolutely fascinating proposal is that the fabric of spacetime is knitted together by the quantum entanglement of whatever the underlying 'atoms' of spacetime are," Balasubramanian says. "That's amazing if true."

The reasoning behind the idea comes from several earlier discoveries by physicists, such as a 2006 paper by Shinsei Ryu and Tadashi Takayanagi showing a connection between entanglement and the geometry of spacetime. Building on that work, in 2013 Juan Maldacena and Leonard Susskind found that if two black holes became entangled, they would create a wormhole–a shortcut in spacetime predicted by general relativity. This discovery (nicknamed ER=EPR, after physicists' shorthand for wormholes and entanglement) and others like it suggest, surprisingly, that entanglement–which was thought to involve no physical link–can produce structures in spacetime.

To understand how entanglement might give rise to spacetime, physicists first must better understand how entanglement works. The phenomenon has seemed "spooky," in the words of Albert Einstein, ever since he and collaborators predicted it in 1935. Lately scientists have been studying the various kinds of entanglement that can exist. For instance, conventional entanglement involves linking a single characteristic (such as a particle's spin) in multiple particles of the same type spread out in space. But one could instead entangle multiple particles of a certain kind at one location with particles of a different kind at the same location. "That's not entanglement in space," Balasubramanian says. "I've come to realize that there are other forms of entanglement that turn out to be relevant for this project of reconstructing spacetime—conventional entanglement is not enough." Scientists are also tackling the confusing complexities of entangling larger numbers of particles.

Once the dynamics of entanglement are clearer, scientists hope to comprehend how spacetime emerges, just as the microscopic movements of atoms in the air give rise to the complex patterns of thermodynamics and weather. "This is an emergent phenomenon—when you zoom out of something, you see a different picture that you wouldn't know comes about because of smaller dynamics," Engelhardt says. "This is one of the most fascinating things about It from Qubit, because we don't understand the fundamental quantum dynamics from which spacetime emerges."

Cosmic Holograms

The major goal of all this work is to finally achieve a theory that describes gravity from a quantum perspective. Yet physicists chasing this goal have been stymied for a century so far—Einstein himself pursued such a theory doggedly until his death, with no success. The It from Qubit scientists are banking on an idea known as the holographic principle to help them.

This principle suggests that some physical theories are equivalent to simpler theories that work in a lower-dimensional universe, in the same way that a 2-D postcard with a hologram of a unicorn on it can contain all the information necessary to describe and portray the 3-D shape of the unicorn. Because finding a working theory of quantum gravity is so hard, the thinking goes, physicists could aim to discover an equivalent, easier-to-work-with theory that operates in a universe with fewer dimensions than ours.

One of the most successful embodiments of the holographic principle is a discovery known as the AdS/CFT correspondence, found by Maldacena in 1997 within the framework of string theory. String theory, itself an attempt at a theory of quantum gravity, replaces all the fundamental particles of nature with tiny vibrating strings. In the AdS/CFT correspondence Maldacena showed that one can completely describe a black hole purely by describing what happens on its surface. In other words, the physics of the inside–the 3-D "bulk"–corresponds perfectly to the physics of the outside–the 2-D "boundary."

AdS/CFT might allow physicists to discover a theory that is equivalent to quantum gravity, accomplishes all the same goals and can describe all the same physics, but that is much easier to work with–by leaving out gravity altogether. "Theories with gravity are very difficult to get quantum descriptions of whereas theories that don't have gravity are much easier to describe completely," Balasubramanian says. But how, one might ask, could a theory that leaves out gravity ever be a theory of "quantum gravity"? Perhaps what we think of as gravity and spacetime is just another way of looking at the end product of entanglement–in other words, entanglement might somehow encode the information from the 3-D bulk into bits stored on the 2-D boundary. "It's a very exciting direction," he adds.

For the past 20 years scientists have found that the AdS/CFT correspondence works–a 2-D theory can describe a 3-D situation–but they do not fully understand why. "We know these two theories are dual but it's not exactly clear what makes the duality work," Swingle says. "One output [of IfQ] you could hope for is a theory for how these dualities arise. That's something I think definitely can and will happen as a result of this collaboration or at least major progress towards that."

Quantum information theory may be able to help because it turns out that a familiar concept from this field, quantum error–correcting codes, could be at work in the AdS/CFT correspondence. In quantum computers, quantum error–correcting codes are a method scientists devised to help protect information from being lost if the entanglement between any particular bits gets broken. Rather than using single bits to encode information, quantum computers use highly entangled states of multiple bits to stand in for each bit, so that a single error cannot affect the overall bit. "There's an underlying mathematical structure that seems to be common to the error-correcting codes and AdS/CFT," says quantum information scientist Dorit Aharonov, an IfQ principal investigator at The Hebrew University of Jerusalem. In computers that redundancy is being used to correct errors, but in AdS/CFT it may be able to encode the bulk physics into an entangled state on the boundary. "It's very intriguing that you find quantum error–correcting codes inside black holes," she says. "Why on Earth would that happen? These connections are just fascinating."

If physicists do eventually understand how the AdS/CFT correspondence works–and come up with a lower-dimensional theory that stands in for quantum gravity–they are still not home free. The correspondence itself only works in a "toy model" of the universe that is somewhat simplified from the fully realized cosmos we inhabit. "AdS/CFT has a kind of gravity, but it's not the theory of gravity in an expanding universe like we live in," Swingle says. "It describes a universe as if it was in a bottle–if you shine a light beam, it bounces off the walls of the space. That doesn't happen in our expanding universe." This model gives physicists a useful theoretical playground in which to test their ideas, where the simplified picture makes tackling quantum gravity easier. "You can hope it's a useful way station in the eventual goal of understanding gravity in our own universe," Swingle explains.

Some skeptics have questioned how productive IfQ can ever be if it is based on an unrealistic foundation. "That certainly is one very valid criticism: Why are we focusing on this toy model?" Engelhardt says. "All of this depends on the validity of the toy model, and the idea that in the end the toy model is representative of our universe. I would like to make sure that if we understand the toy model, we understand the real deal."

The Payoff

Regardless of whether It from Qubit will ultimately achieve the holy grail of a unified theory, scientists inside and outside the project say the approach is worth trying and is already turning up many new avenues to pursue. "I've long felt that the relation between quantum information and quantum gravity is of fundamental importance," says Raphael Bousso, a physicist at the University of California, Berkeley, who is not involved in IfQ but has worked with some of its collaborators. "The connection has deepened over the years, and I'm thrilled that so many outstanding scientists are now working together to confront these questions and see where they lead us." Stanford University theorist Eva Silverstein, who is not part of the collaboration, concurs. "It is clearly worthwhile to develop and apply quantum information to these problems. But to understand the dynamics [of quantum gravity] much more is required, and it is important for the field not to focus too narrowly on a single approach."

Furthermore, even if the project does not pay off with a theory of quantum gravity, it is still likely to have beneficial offshoots. Bringing the techniques and ideas of string theory and general relativity to bear on questions of quantum information can, for instance, help to better define the different types of entanglement that can exist, both for purposes of understanding spacetime as well as constructing quantum computers. "When you start playing with these tools in a new setting, it's very likely that it will bring up ideas that are interesting and might be useful in other areas," Aharonov says. "It looks like people are making progress on questions that have been out there for many, many years, so it's exciting." For instance, scientists have found that measuring time within wormholes may be possible by thinking of the wormhole as a quantum circuit.

Furthermore, combining quantum information science with string theory may help not just in deriving a theory of quantum gravity but in evaluating whatever theory the researchers find. "A crucial question we would ask—once we actually manage to come up with a detailed-enough physical theory of quantum gravity—is what is the computational power of this model?" Aharonov says. Any physical theory can be thought of as a computational model, its input and output akin to the theory's initial state and a later state that can be measured—and any computational model has a computational power. "If that power is too large, if our quantum gravity model would be able to compute things that we don't believe can be computed in our world, that would at least raise a question mark on the theory. It's a way to actually tell whether the theory is sensible or not from a different point of view."

The project is reminding some physicists of the heady days in the past when other big ideas were just getting started. "I became a grad student in 1984 when the so-called 'first string theory revolution' took place," says Hirosi Ooguri, a physicist at the California Institute of Technology who has been working on IfQ. "That was a very exciting time when string theory emerged as a leading candidate for a unified theory of all the forces in nature. I do see the current explosion of excitement around this similarly. This is clearly an exciting time for young people in the field as well as those of us who received our PhDs decades ago."

About the Author

Clara Moskowitz is a senior editor at Scientific American, *where she covers space and physics.*

Found: A Quadrillion Ways for String Theory to Make Our Universe

By Anil Ananthaswamy

Physicists who have been roaming the "landscape" of string theory–the space of zillions and zillions of mathematical solutions of the theory, where each solution provides the kinds of equations physicists need to describe reality–have stumbled upon a subset of such equations that have the same set of matter particles as exists in our universe.

But this is no small subset: there are at least a quadrillion such solutions, making it the largest such set ever found in string theory.

According to string theory, all particles and fundamental forces arise from the vibrational states of tiny strings. For mathematical consistency, these strings vibrate in 10-dimensional spacetime. And for consistency with our familiar everyday experience of the universe, with three spatial dimensions and the dimension of time, the additional six dimensions are "compactified" so as to be undetectable.

Different compactifications lead to different solutions. In string theory, a "solution" implies a vacuum of spacetime that is governed by Einstein's theory of gravity coupled to a quantum field theory. Each solution describes a unique universe, with its own set of particles, fundamental forces and other such defining properties.

Some string theorists have focused their efforts on trying to find ways to connect string theory to properties of our known, observable universe–particularly the standard model of particle physics, which describes all known particles and all their mutual forces except gravity.

Much of this effort has involved a version of string theory in which the strings interact weakly. However, in the past two decades, a new branch of string theory called F-theory has allowed physicists to work with strongly interacting, or strongly coupled, strings. "An intriguing, surprising result is that when the coupling is large, we can start describing the theory very geometrically," says Mirjam Cvetic of the University of Pennsylvania in Philadelphia.

This means that string theorists can use algebraic geometry–which uses algebraic techniques to tackle geometric problems–to analyze the various ways of compactifying extra dimensions in F-theory and to find solutions. Mathematicians have been independently studying some of the geometric forms that appear in F-theory. "They provide us physicists a vast toolkit," says Ling Lin, also of the University of Pennsylvania. "The geometry is really the key... it is the 'language' that makes F-theory such a powerful framework."

Now, Cvetic, Lin, James Halverson of Northeastern University in Boston, and their colleagues have used such techniques to identify a class of solutions with string vibrational modes that lead to a similar spectrum of fermions (or, particles of matter) as is described by the standard model–including the property that all fermions come in three generations (for example, the electron, muon and tau are the three generations of one type of fermion).

The F-theory solutions found by Cvetic and colleagues have particles that also exhibit the handedness, or chirality, of the standard model particles. In particle physics lingo, the solutions reproduce the exact "chiral spectrum" of standard model particles. For example, the quarks and leptons in these solutions come in left and right-handed versions, as they do in our universe.

The new work shows that there are at least a quadrillion solutions in which particles have the same chiral spectrum as the standard model, which is 10 orders of magnitude more solutions than had been found within string theory until now. "This is by far the largest domain of standard model solutions," Cvetic says. "It's somehow surprising and actually also rewarding that it turns out to be in the strongly coupled string theory regime, where geometry helped us."

A quadrillion–while it's much, much smaller than the size of the landscape of solutions in F-theory (which at last count was shown to be of the order of $10^{272,000}$)–is a tremendously large number. "And because it's a tremendously large number, and it gets something nontrivial in real world particle physics correct, we should take it seriously and study it further," Halverson says.

Further study would involve uncovering stronger connections with the particle physics of the real world. The researchers still have to work out the couplings or interactions between particles in the F-theory solutions–which again depend on the geometric details of the compactifications of the extra dimensions.

It could be that within the space of the quadrillion solutions, there are some with couplings that could cause the proton to decay within observable timescales. This would clearly be at odds with the real world, as experiments have yet to see any sign of protons decaying. Alternatively, physicists could search for solutions that realize the spectrum of standard model particles that preserve a mathematical symmetry called R-parity. "This symmetry forbids certain proton decay processes and would be very attractive from a particle physics point of view, but is missing in our current models," Lin says.

Also, the work assumes supersymmetry, which means that all the standard model particles have partner particles. String theory needs this symmetry in order to ensure the mathematical consistency of solutions.

But in order for any supersymmetric theory to tally with the observable universe, the symmetry has to be broken (much like how a diner's selection of cutlery and drinking glass on her left or right side will "break" the symmetry of the table setting at a round dinner table). Else, the partner particles would have the same mass as standard model particles–and that is clearly not the case, since we don't observe any such partner particles in our experiments.

Crucially, experiments at the Large Hadron Collider (LHC) have also shown that supersymmetry–if it is the correct description of nature–is not broken even at the energy scales probed by the LHC, given that the LHC has yet to find any supersymmetric particles.

String theorists think that supersymmetry might be broken only at extremely high energies that are not within experimental reach anytime soon. "The expectation in string theory is that high-scale [supersymmetry] breaking, which is fully consistent with LHC data, is completely possible," Halverson says. "It requires further analysis to determine whether or not it happens in our case."

Despite these caveats, other string theorists are approving of the new work. "This is definitely a step forward in demonstrating that string theory gives rise to many solutions with features of the standard model," says string theorist Washington Taylor of MIT.

"It's very nice work," says Cumrun Vafa, one of the developers of F-theory, at Harvard University. "The fact you can arrange the geometry and topology to fit with not only Einstein's equations, but also with the [particle] spectrum that we want, is not trivial. It works out nicely here."

But Vafa and Taylor both caution that these solutions are far from matching perfectly with the standard model. Getting solutions to match exactly with the particle physics of our world is one of the ultimate goals of string theory. Vafa is among those who think that, despite the immensity of the landscape of solutions, there exists a unique solution that matches our universe. "I bet there is exactly one," he says. But, "to pinpoint this is not going to be easy."

About the Author

Anil Ananthaswamy is author of The Edge of Physics *(Houghton Mifflin Harcourt, 2010),* The Man Who Wasn't There *(Dutton, 2015), and* Through Two Doors at Once: The Elegant Experiment That Captures the Enigma of Our Quantum Reality *(Dutton, 2018).*

Quantum Entanglement Isn't All That Spooky After All

By Chris Ferrie

Quantum entanglement is a complex phenomenon in physics that is usually poorly described as an invisible link between distant quantum objects that allows one to instantly affect the other. Albert Einstein famously dismissed this idea of entanglement as "spooky action at a distance." In reality, entanglement is better understood as information, but that's admittedly bland. So nowadays, every news article, explainer, opinion piece and artistic interpretation of quantum entanglement equates the phenomenon with Einstein's spookiness. The situation has only worsened with the 2022 Nobel Prize in Physics going to Alain Aspect, John F. Clauser and Anton Zeilinger for quantum entanglement experiments. But it's time to cut this adjective loose. Calling entanglement spooky completely misrepresents how it actually works and hinders our ability to make sense of it.

In 1935, physicist Erwin Schrödinger coined the term entanglement, emphasizing that it was "not *one* but rather *the* characteristic trait of quantum mechanics, the one that enforces its entire departure from classical lines of thought." He was writing in response to a famous paper (known simply to physicists as the EPR argument) by Einstein, Boris Podolsky and Nathan Rosen, that claimed quantum physics was incomplete. The *New York Times* headline read, "Einstein attacks quantum theory," which solidified the widespread perception that Einstein hated quantum physics.

The EPR argument concerns the everyday notion of reality as a collection of things in the world with physical properties waiting to be revealed through measurement. This is how most of us intuitively understand reality. Einstein's theory of relativity fits into this understanding, and says reality must be *local*, meaning nothing can influence anything else faster than the speed of light.

But EPR showed that quantum physics isn't compatible with these ideas–that it can't account for a theory of local reality. In other words, quantum physics was missing something. To complete quantum physics, Einstein suggested scientists should look for a "deeper" theory of local reality. Many physicists responded in defense of quantum theory, but the matter remained unresolved until 1964 when physicist John S. Bell proposed an experiment that could rule out the existence of a local reality. Clauser was the first to perform the test, which was later improved and perfected by Aspect and Zeilinger.

A typical article about entanglement tells us it arises when particles interact to create a "link," which persists no matter how far apart those particles are. Moreover, actions taken on one particle instantly affect the other, or so we are told. But–and here's the thing even many experts get wrong–quantum physics *doesn't* say that. Quantum physics says nothing about how the world is. Instead, quantum physics only describes the experiments we do to test our theories of how the world works–it gives us probabilities for the outcomes that may happen in an experiment. The compulsion to interpret quantum physics concepts as prescriptions for physical reality derives from the unfortunate way we traditionally teach physics.

I teach quantum physics to second-year computer science students at the University of Technology Sydney. Every autumn, I give teenagers a working knowledge of quantum entanglement without telling them it is spooky by guiding them through the process of engineering quantum phenomena for themselves. A former student said they understood the 2022 Nobel Prize in Physics reporting because I have students program quantum computers to produce entanglement. Another former student told me they were having trouble figuring out where the mysterious spookiness was supposed to be. I suggested that perhaps they needn't look for something they're not going to find.

Typically, a physics teacher starts a lecture on entanglement with Einstein, introducing concepts like local realism and ending with necessarily invoking the free will of the experimenter. But it doesn't need to be this way. It's much easier to understand how quantum physics works, and how it departs from the classical world, from the perspective of information, not physics. Let's consider an example.

Imagine two people, Alice and Bob, are implicated in a crime and are being questioned in separate rooms with no way to communicate. They are each asked one of two possible questions. They must corroborate each other's story to be set free. But there's a catch: the questions contain a trap such that if they are both asked the second question, they must give *opposite* answers. Alice and Bob know all this before heading into their rooms for questioning. So, they do the obvious thing and devise a strategy so that their answers will be correlated in just the right way. However, it quickly becomes apparent that no possible strategy can set them free since they won't know which question the other investigator asked. The best Alice and Bob can do is answer correctly 75 percent of the time, by both giving the same answer for every question, accepting they will fail in one of the four cases.

So far, Alice and Bob have only used classical information. But by sharing quantum information, they succeed with a probability higher than 75 percent. They do this by devising a strategy using the mathematics of quantum information rather than classical information. Intuiting the solution requires some familiarity with linear algebra, so I won't detail it here. But it is a fact that the quantum information they share requires correlations, which means it is entangled. This appears spooky to the investigators because they only reason with classical information. But it's not spooky. In any theory of information, correlations are ubiquitous. Through the lens of quantum information, then, entanglement is not strange or rare, but rather expected. The information perspective beautifully illustrates the core problem with demanding a classical description of quantum phenomena: it's the wrong language. The Nobel Prize–winners were the first to demonstrate this as a fact about nature. Today, you can follow in their footsteps by creating entanglement and processing the correlated quantum information on a real quantum computer.

Einstein wanted all of nature identified with a simple and compact classical description. But we now know that quantum information provides the most accurate description of nature, which is written in a language we do not speak. Accepting this liberates us from the limits of traditional physics and makes teaching it more natural by facilitating active learning. The quantum information perspective illuminates some of the most profound questions in physics. For example, quantum information is the key to understanding the mystery of black holes and perhaps the entire universe. It also leads us to new quantum technologies that quickly and automatically encode and process quantum information.

For the second half of the 20th century, computers rapidly changed every facet of society, transforming our understanding of the universe and ourselves. We thought they were the ultimate tool for this purpose, but we were wrong. Scientists now believe the ultimate machine is a quantum computer, the full potential of which we have yet to realize. Determining when quantum computers will become ubiquitous and what problems they will solve is an exercise in crystal gazing. However, we already know they can solve a small list of problems such as factoring numbers, searching databases or simulating chemical reactions. If you have such a problem, you might be in the market for a quantum computer. You'll love it; Einstein, on the other hand, would have hated it.

About the Author

Chris Ferrie is a faculty member of the Center for Quantum Software and Information at the University of Technology Sydney. He is the author of Quantum Physics for Babies *and* Quantum Bullsh*t: How to Ruin Your Life With Advice from Quantum Physics.

Section 5: Gravitational Waves

Tiny Gravitational-Wave Detector Could Search Anywhere in the Sky

By Tim Folger

The smallest, most precise measurement ever made required one of the largest scientific instruments ever constructed. Five years ago the Laser Interferometer Gravitational-Wave Observatory (LIGO) detected a ripple in spacetime that was just one ten-thousandth the width of a proton—a technical tour de force tantamount to pinpointing the distance to the nearest star to three one-thousandths of a centimeter. The Lilliputian ripple was a gravitational wave, a distortion in the fabric of the cosmos generated by the collision of two black holes more than a billion light-years from Earth.

According to Albert Einstein's general theory of relativity, the acceleration of any massive object creates waves in spacetime, much as a ship churns out waves in water. Einstein himself, however, thought gravitational waves would be too weak to detect. He was not unduly pessimistic. It took LIGO's four-kilometer-long interferometers—which were completed in 1999 and began searching for the waves in 2001—13 years to finally spot one. The discovery marked the beginning of a new field of astronomy and netted a Nobel Prize for three of the observatory's physicists. The experiment has since detected nearly a dozen more gravitational-wave events. Now, just as LIGO is hitting its stride, a team of physicists has outlined a way to build a portable gravitational detector that is only one meter long—4,000 times smaller than LIGO.

The proposal, which will be published soon in *New Journal of Physics*, describes a detector that would exploit a strange quantum mechanical phenomenon to reveal the passage of a gravitational wave. "The first and most important thing to know from the experimental side is that it would be phenomenally difficult to build it," says Gavin Morley, a physicist at the University of Warwick in England and one of the co-authors of the study. If the team succeeds, though, the new device would offer a much more compact way of detecting gravitational waves that could be reproduced in many laboratories around the world.

Waves Canceling Out

The physicists call their proposed device the Mesoscopic Interference for Metric and Curvature (MIMAC). Despite their enormous size disparities, both MIMAC and LIGO look for the same effect: the rhythmic stretching and contraction of spacetime caused by a gravitational wave traveling at the speed of light.

In LIGO's case, two identical instruments—one in Livingston, La., and the other in Hanford, Wash.—were built to rule out false signals from local gravitational effects. Each site has two four-kilometer-long vacuum chambers that meet at a 90-degree angle, forming a big L on the landscape. Forty-kilogram mirrors made of highly pure silica sit at both ends of each vacuum chamber. A laser beam constantly flits back and forth between the mirrors, monitored by a light detector at the corner of the L.

LIGO was designed so that under normal conditions, the light waves from each arm cancel out when they meet at the detector: no signal reaches it because the crests and troughs from each arm's light overlap. But if a gravitational wave passes through the arms, it periodically stretches one of them and compresses the other, changing their length by a fraction of a proton's diameter. Then the light waves no longer cancel each other: they send light pulses to the detector in sync with the passing gravitational wave, creating a distinctive flickering pattern.

So how could a meter-long device possibly accomplish the same feat? A key component of MIMAC would be a diamond particle no larger than a millionth of a meter. The researchers want to put such a diamond into a quantum superposition—a state in which the diamond would occupy two different positions simultaneously—and then wait for it to interact with a gravitational wave.

A Flawed Diamond

In addition to Morley, the team includes Sougato Bose, Peter Barker and Ryan Marshman, all at University College London, along with Anupam Mazumdar and Steven Hoekstra, both at the University of Groningen in the Netherlands. To create the superposition, they would beam microwaves at a single electron bound to a manufactured flaw in the diamond's crystal lattice of carbon atoms. (The flaw consists of a single nitrogen atom inserted into the otherwise uniform array of carbon.) Then the extraordinary rules of quantum theory would kick in: the electron would, at once, both absorb and not absorb a microwave photon, creating a quantum superposition of the diamond. The electron in the diamond doppelgänger that absorbed the photon would shift to a so-called "spin one" state, meaning it behaves like a miniature magnet with its own magnetic field. The electron in the other version of the diamond would remain in a "spin zero" state–magnetically neutral. By applying an external magnetic field, Bose and his colleagues say it should be possible to pull the spin-one part of the superposition away from its neutral counterpart, separating them by as much as a meter. Finally, the physicists would reverse the magnetic field, bringing the two positions of the diamond back together, and hit it with one last microwave pulse.

That last pulse would trigger another strange quantum effect. In the quantum realm, particles are not really particles per se. They are actually waves, and their shape and size corresponds to the probability of finding a "particle" at a given position. The final burst of microwaves would be tuned to change the shape of the superposition so that the crests and troughs of the spin-one state overlap and cancel out, while the crests of the spin-zero state overlap and reinforce each other. Thus, in the absence of any outside interference, a measurement of the electron would always find it in a spin-zero state.

But any gravitational wave surging over the detector would stretch the superposition, changing its shape so that its components no longer aligned when rejoined. Measurements of the distorted superposition would then yield mixed results, with the spin-one state turning up in the data in sync with the frequency of the gravitational wave.

That scenario is the theory at least. Building a working model may take decades. Ron Folman, an experimental physicist at Ben-Gurion University of the Negev in Israel, who was not involved in the proposal, calls the idea "audacious." Isolating the system so that the quantum particles do not interact with the environment will be extremely challenging, he says. "It's a very hard experiment," he adds, but it "may be realized within our lifetime, given enough dedicated effort."

One of the biggest challenges will be to create superpositions of diamonds that can remain stable over distances of a meter. More than four years ago researchers at Stanford University managed to separate a superposition consisting of 10,000 atoms by about half a meter–the current record. "But we're talking about doing it with diamonds that would have a billion or 10 billion atoms, and that is way more difficult," Mazumdar says.

Many of the other technologies needed for the device–high vacuums, ultralow temperatures, precisely controlled magnetic fields–have all been achieved separately by various groups. But bringing them together will not be easy. "Just because you can juggle and ride a bike doesn't mean you can do both at once," Morley says.

If the device is ever built, it could transform gravitational-wave astronomy. The world's current gravitational-wave detectors are all firmly anchored to the ground. "The only orientation LIGO can have is due to Earth's rotation," Bose says. A small detector such as MIMAC, on the other hand, could be pointed at any direction in the sky. And any physics lab in the world could house it. "The challenge is to get one of them working," Bose says. "If one of them works, it would be very easy to make several more."

About the Author

Tim Folger is a freelance journalist who writes for National Geographic, Discover, *and other national publications.*

Gravitational Waves Probe Exotic Matter inside Neutron Stars

By Clara Moskowitz

A mystery lurks inside the corpses of dead stars. Neutron stars, formed when certain types of stars die in supernova explosions, are the densest form of matter in the universe; black holes are the only thing denser, but they have so fully escaped the bounds of normal physics that they are not matter anymore. The atoms in neutron stars have been squeezed so tightly by gravity that they have broken down, the protons and electrons inside them smushing together to create neutrons, leaving objects the size of small cities that contain masses larger than the sun. Some 95 percent of the mass of a neutron star is pure neutrons, but physicists wonder what happens at the very center, where the density peaks. Do neutrons break down further into their constituent quarks and gluons? Do some of the quarks transform from their normal "up" and "down" flavors to become weirder and heavier "strange quarks" not found in ordinary matter? Do the particles form an extreme state of matter called a superfluid that sloshes with no viscosity, never slowing down?

Scientists have come a step closer to understanding the inner workings of these bizarre bodies by studying the light and gravitational waves that result when two neutron stars slam into each other and become a black hole. Gravitational waves are folds in spacetime carved out when large masses move around. Scientists only gained the ability to detect gravitational waves in 2015, and have spotted just a handful of events involving neutron stars so far (the others have been collisions of black holes). But studying the properties of these waves–their frequency and how they change over time–can tell scientists a lot about the objects that created them. Physicists seek accurate measurements of neutron stars' masses and radii, which would help reveal their "equation of state"–the relationship between pressure and density within these stars.

Knowing the neutron star equation of state would in turn indicate what kind of matter hides inside them.

In a new study, an international team of researchers combined gravitational wave measurements from two neutron star collisions, as well as the light signals that arrived along with one of them (the other was dark), with estimates of neutron star masses and radii from watching rapidly spinning neutron stars called pulsars. "The big advantage is it's a very coherent picture," says study member Tim Dietrich of the University of Potsdam in Germany, who co-authored a paper reporting the results published today in *Science*. "We combine all the things we know currently, including gravitational waves and electromagnetic waves, information from single neutron stars, and theoretical computations from nuclear physics." The equation of state they derived predicted that a neutron star containing the mass of 1.4 suns would have a radius of about 11.75 kilometers, plus or minus .81 to .86 km. That's a bit more than half the length of Manhattan. "The size of the neutron star directly depends on the behavior of matter inside the core, so this gives us a better understanding about the properties of the neutron star material," Dietrich says.

For instance, if neutrons remain intact in the core of these stars, they would push out against the outer layers, potentially leading to a slightly larger radius. If, on the other hand, the neutrons break down into a soup of quarks, the core would be squishier and the whole star would sink in a bit, resulting in a smaller radius.

The new measurement is in general agreement with earlier studies that have looked at gravitational wave data and other ways to measure neutron star size. "This paper is a nice joint reanalysis of previous studies, and doesn't change the overall impression that has been in place for the last few years, that the radius of a neutron star is about 11 to 13 km," says Mark Alford, a physicist at Washington University in St. Louis. Anna Watts, an astrophysicist at the University of Amsterdam, says this type of combined analysis "is clearly the way forward," but that none of the measurements "are yet good enough to really pin down the nature of dense matter." The field will need to wait for future data to really understand what's happening inside neutron stars.

"I think it's a very nice analysis," says physicist James Lattimer of Stony Brook University, who was not involved in the research. He cautions that in modeling how well different possible equations of state fit the data, the team may have mistakenly eliminated too many equations that produce neutron stars with large radii, however. "I think they've underestimated their uncertainty. But in some sense, it's a matter of opinion and how much faith you place in different statistical methods."

Besides revealing secrets of neutron stars, the study also produced a measurement of the Hubble constant, which reflects the expansion rate of the universe. To derive the constant, the scientists used the amplitude of the gravitational waves coming from one of the collisions to estimate how far away the crash occurred. They then compared their distance measurement with the known speed of the collision's host galaxy, which was measured by looking at the galaxy's redshift–how much its light has slid toward the red end of the spectrum. The Hubble constant they found, 66.2 kilometers per second per megaparsec, is not precise enough to decide between the competing measurements that already exist, but adds another data point to the hotly contested question of how fast the cosmos is growing.

The scientists hope to apply the same type of analysis to future neutron star collisions that appear. "We made this first step and now we'll push forward," says team member Sarah Antier of the University of Paris, an astronomer who searches for light signals accompanying gravitational wave events. "My task is to connect different observatories to provide a network to make immediate observations" when gravitational wave detectors find a new signal.

Physicists are biding their time until the next generation of gravitational-wave detectors, such as the Cosmic Explorer in the U.S. and the Einstein Telescope in Europe, come online in the 2030s. These machines should be much more sensitive, allowing them to capture many more signals from more events and offering higher precision data. Future projects such as the Enhanced X-ray Timing and Polarimetry Mission (eXTP) and the Athena X-ray Observatory should also gather more accurate measurements of pulsars.

Scientists have learned so much in the short time since gravitational wave data became available, the future promises to greatly expand our knowledge of extreme matter under intense pressure. "The last four years have been remarkable," Lattimer says. "It shows the potential that we are going to be getting in the future. We should have many more measurements from gravitational wave events, and as we add each new event the results are going to converge."

About the Author

Clara Moskowitz is a senior editor at Scientific American, *where she covers space and physics.*

Galaxy-Size Gravitational-Wave Detector Hints at Exotic Physics

By Adam Mann

The fabric of spacetime may be frothing with gigantic gravitational waves, and the possibility has sent physicists into a tizzy. A potential signal seen in the light from dead stellar cores known as pulsars has driven a flurry of theoretical papers speculating about exotic explanations.

The most mundane, yet still quite sensational, possibility is that researchers working with the North American Nanohertz Observatory for Gravitational Waves (NANOGrav), which uses the galaxy as a colossal gravitational-wave detector, have finally seen a long-sought background signature produced when supermassive black holes crash and merge throughout the universe. Another interpretation would have it originating from a vibrating network of high-energy cosmic strings that could provide scientists with extremely detailed information about the fundamental constituents of physical reality. A third possibility posits that the collaboration has spotted the creation of countless small black holes at the dawn of time, which could themselves account for the mysterious substance known as dark matter.

"People have been making predictions about cosmic strings and primordial black holes for years, and now, finally, we have a signal," says Chiara Mingarelli, an astrophysicist at the University of Connecticut and a member of the NANOGrav team. "We're not sure what is generating this signal, but a lot of people are really, really excited."

The physics community has learned a great deal about the universe from massive terrestrial gravitational wave experiments such as the Laser Interferometer Gravitational-Wave Observatory (LIGO) and its European counterpart Virgo. But just as electromagnetic waves come in a spectrum ranging from squashed gamma-rays to lengthy radio waves, gravitational waves run the gamut from the tiny vibrations in spacetime made when sun-size black holes merge to those with wavelengths measurable in light-years that can take decades to pass by our planet. The collective, overlapping cacophonies from those larger waves, thought to be produced when behemoth black holes lurking in the centers of galaxies collide, are what the NANOGrav collaboration has been working to capture.

It does so by focusing on objects known as millisecond pulsars, which arise when massive stars explode as supernovae and leave behind their rapidly spinning remnant hearts. A pulsar's strong magnetic field can create a beam of radiation that swings around, repeatedly sweeping past Earth with a regularity that rivals the accuracy of atomic clocks. Should a distortion in the fabric of spacetime come between our planet and a pulsar, it can cause this signal to arrive slightly earlier or later than expected. Were a telescope to see one such offset, it probably would not mean much. But NANOGrav has been monitoring the light from 45 pulsars scattered over thousands of light-years for more than 12.5 years, looking for correlations between their arrival times that could indicate the presence of gravitational waves.

Last September, the collaboration posted a paper on the preprint server arXiv.org, which hosts scientific articles that have yet to go through peer review, showing that its monitored pulsars all displayed similar blips. (The paper has since been peer-reviewed and published.) The chances of this happening are between 1,000 and 10,000 to one, say Mingarelli. As a group, NANOGrav is cautious and has refrained from claiming it has seen a gravitational-wave signal, which requires observing highly specific correlations among its pulsar signals' arrival times. That did not stop other scientists from jumping on the data.

Marek Lewicki, a theoretical physicist at the University of Warsaw in Poland, recalls that the NANOGrav study appeared early on a Friday morning and that, by 10 A.M., his collaborator John Ellis of King's College London had spotted it. Though the usual explanation for such a signal is the supermassive black hole gravitational-wave background, Lewicki knew that another possible culprit was cosmic strings, and he began running models to see if this option could account for the data. "By Saturday, it was pretty clear it was a good fit," he says.

Researchers like cosmic strings because they directly connect cosmological events to high-energy particle physics. Shortly after the big bang, three of the four known forces—electromagnetism and the strong and weak nuclear forces—would have been smushed together into one superforce. When the strong nuclear force dissociated itself, the universe would have gone through what is known as a phase change, much like water freezing into ice. And just as a frozen lake often contains long cracks created when its bulk solidifies, the visible cosmos would become strewn with enormous nearly-one-dimensional tubes of energy crisscrossing its length. Such objects would be tense like piano strings and could vibrate out gravitational waves that would look like the signal NANOGrav had picked up.

Because these cosmic strings originated near the beginning of time, they would carry information about processes such as cosmic inflation, during which the universe is thought to have rapidly ballooned by mind-boggling factors, as well as the creation of different particles at different extreme temperatures, says Kai Schmitz, a theoretical physicist at CERN near Geneva. Information from such conditions, which would be impossible to create in particle accelerators such as the Large Hadron Collider, could help researchers produce a grand unified theory connecting most known particles and forces that would supersede the current Standard Model. Along with two collaborators, Schmitz published a paper in *Physical Review Letters* (*PRL*) outlining how cosmic strings could account for the NANOGrav data on January 28, the same day a similar article by Lewicki and Ellis appeared.

"If we detected cosmic strings, it would be the detection of my lifetime," says Eugene Lim, a cosmologist also at King's College London. "It would be more important than the Higgs boson, probably more than gravitational waves themselves."

For this reason, Lim, who was not a co-author on either paper, stresses that such concepts need to be considered with an abundance of restraint. The NANOGrav collaboration still needs to confirm that it is in fact seeing gravitational waves. And the shape of those gravitational waves' spectrum has yet to be traced out and found to conform to the cosmic string interpretation, each of which is likely to take years, he adds.

Meanwhile, another contingent of the physics community has suggested that the signal could originate from entities known as primordial black holes. Unlike regular black holes, which are born when gigantic stars die, these would form in the early universe, when matter and energy were nonuniformly scattered through the cosmos as a consequence of processes that occurred at the end of inflation. Certain overdense areas could collapse under their own weight, generating black holes in a variety of sizes. Observations from LIGO and Virgo that could indicate mergers between primordial black holes have already planted the idea in many researchers' minds that these strange objects are more than speculative fictions. Certain theorists like them because, as entities that give off no light, they could account for some or even all of the dark matter in the universe.

"This is an economical explanation," says Antonio Riotto, an astroparticle cosmologist at the University of Geneva, because they do not require theorizing about exotic undetected particles such as WIMPs or axions, which have thus far dominated physicists' musings about dark matter.

Along with two co-authors, Riotto has written a third paper appearing in *PRL* showing how the NANOGrav signal could be accounted for by a multitude of black holes the size of asteroids being created shortly after the big bang, producing a gravitational wave relic that would travel to us in the modern day. According to the researchers' model, these miniature primordial black holes could comprise up to 100 percent of the dark matter in the universe.

Yet this possibility, too, needs to be approached carefully, says Juan García-Bellido, a theoretical physicist at the Autonomous University of Madrid in Spain, who was not involved in the work.

While the NANOGrav data contain hints, it does not quite show the specific correlated pattern that would indicate gravitational waves, and much of the speculation seems premature to him. "I'm the first to hope for primordial black holes," he says. "But I'm afraid it's not yet there."

Nevertheless, the burst of theoretical activity shows how seriously physicists are taking these results. NANOGrav researchers have another two and a half years of pulsar data they are combing through, which could help distinguish whether some or a combination of all these explanations might be viable. They are also working with international collaborators such as the European Pulsar Timing Array (EPTA) and Parkes Pulsar Timing Array (PPTA) in Australia, each of which has observations of other pulsars that could get them closer to spotting the needed correlations to finally pin down the gravitational-wave background—a process that should be underway before the end of this year.

"I would be shocked if we didn't see a signal when we combined all of our data," Mingarelli says.

About the Author

Adam Mann is a journalist specializing in astronomy and physics. His work has appeared in National Geographic, *the* Wall Street Journal, Wired, *and elsewhere.*

Section 6: Ongoing Mysteries and Unanswered Questions

The Elusive Theory of Everything

By Stephen Hawking, Leonard Mlodinow

A few years ago the city council of Monza, Italy, barred pet owners from keeping goldfish in curved fishbowls. The sponsors of the measure explained that it is cruel to keep a fish in a bowl because the curved sides give the fish a distorted view of reality. Aside from the measure's significance to the poor goldfish, the story raises an interesting philosophical question: How do we know that the reality we perceive is true?

The goldfish is seeing a version of reality that is different from ours, but can we be sure it is any less real? For all we know, we, too, may spend our entire lives staring out at the world through a distorting lens.

In physics, the question is not academic. Indeed, physicists and cosmologists are finding themselves in a similar predicament to the goldfish's. For decades we have strived to come up with an ultimate theory of everything–one complete and consistent set of fundamental laws of nature that explain every aspect of reality. It now appears that this quest may yield not a single theory but a family of interconnected theories, each describing its own version of reality, as if it viewed the universe through its own fishbowl.

This notion may be difficult for many people, including some working scientists, to accept. Most people believe that there is an objective reality out there and that our senses and our science directly convey information about the material world. Classical science is based on the belief that an external world exists whose properties are definite and independent of the observer who perceives them. In philosophy, that belief is called realism.

Those who remember Timothy Leary and the 1960s, however, know of another possibility: one's concept of reality can depend on the mind of the perceiver. That viewpoint, with various subtle differences, goes by names such as antirealism, instrumentalism or idealism. According to those doctrines, the world we know is constructed by the human mind employing sensory data as its raw material and is shaped by the interpretive structure of our brains. This viewpoint may be hard to accept, but it is not difficult to understand. There is no way to remove the observer—us—from our perception of the world.

The way physics has been going, realism is becoming difficult to defend. In classical physics—the physics of Newton that so accurately describes our everyday experience—the interpretation of terms such as object and position is for the most part in harmony with our commonsense, "realistic" understanding of those concepts. As measuring devices, however, we are crude instruments. Physicists have found that everyday objects and the light we see them by are made from objects—such as electrons and photons—that we do not perceive directly. These objects are governed not by classical physics but by the laws of quantum theory.

The reality of quantum theory is a radical departure from that of classical physics. In the framework of quantum theory, particles have neither definite positions nor definite velocities unless and until an observer measures those quantities. In some cases, individual objects do not even have an independent existence but rather exist only as part of an ensemble of many. Quantum physics also has important implications for our concept of the past. In classical physics, the past is assumed to exist as a definite series of events, but according to quantum physics, the past, like the future, is indefinite and exists only as a spectrum of possibilities. Even the universe as a whole has no single past or history. So quantum physics implies a different reality than that of classical physics—even though the latter is consistent with our intuition and still serves us well when we design things such as buildings and bridges.

These examples bring us to a conclusion that provides an important framework with which to interpret modern science. In our view, there is no picture- or theory-independent concept of reality. Instead we adopt a view that we call model-dependent realism: the idea that a physical theory or world picture is a model (generally of a mathematical nature) and a set of rules that connect the elements of the model to observations. According to model-dependent realism, it is pointless to ask whether a model is real, only whether it agrees with observation. If two models agree with observation, neither one can be considered more real than the other. A person can use whichever model is more convenient in the situation under consideration.

Do Not Attempt to Adjust the Picture

The idea of alternative realities is a mainstay of today's popular culture. For example, in the science-fiction film *The Matrix* the human race is unknowingly living in a simulated virtual reality created by intelligent computers to keep them pacified and content while the computers suck their bioelectrical energy (whatever that is). How do we know we are not just computer-generated characters living in a Matrix-like world? If we lived in a synthetic, imaginary world, events would not necessarily have any logic or consistency or obey any laws. The aliens in control might find it more interesting or amusing to see our reactions, for example, if everyone in the world suddenly decided that chocolate was repulsive or that war was not an option, but that has never happened. If the aliens did enforce consistent laws, we would have no way to tell that another reality stood behind the simulated one. It is easy to call the world the aliens live in the "real" one and the computer-generated world a false one. But if—like us—the beings in the simulated world could not gaze into their universe from the outside, they would have no reason to doubt their own pictures of reality.

The goldfish are in a similar situation. Their view is not the same as ours from outside their curved bowl, but they could still formulate scientific laws governing the motion of the objects they observe on the outside. For instance, because light bends as it travels from air to water, a freely moving object that we would observe to move in a straight line would be observed by the goldfish to move along a curved path. The goldfish could formulate scientific laws from their distorted frame of reference that would always hold true and that would enable them to make predictions about the future motion of objects outside the bowl. Their laws would be more complicated than the laws in our frame, but simplicity is a matter of taste. If the goldfish formulated such a theory, we would have to admit the goldfish's view as a valid picture of reality.

A famous real-world example of different pictures of reality is the contrast between Ptolemy's Earth-centered model of the cosmos and Copernicus's sun-centered model. Although it is not uncommon for people to say that Copernicus proved Ptolemy wrong, that is not true. As in the case of our view versus that of the goldfish, one can use either picture as a model of the universe, because we can explain our observations of the heavens by assuming either Earth or the sun to be at rest. Despite its role in philosophical debates over the nature of our universe, the real advantage of the Copernican system is that the equations of motion are much simpler in the frame of reference in which the sun is at rest.

Model-dependent realism applies not only to scientific models but also to the conscious and subconscious mental models we all create to interpret and understand the everyday world. For example, the human brain processes crude data from the optic nerve, combining input from both eyes, enhancing the resolution and filling in gaps such as the one in the retina's blind spot. Moreover, it creates the impression of three-dimensional space from the retina's two-dimensional data. When you see a chair, you have merely used the light scattered by the chair to build a mental image or model of the chair. The brain is so good at model-building that if people are fitted with glasses that turn the images in their eyes upside down, their brains change the model so that they again see things the right way up—hopefully before they try to sit down.

Glimpses of the Deep Theory

In the quest to discover the ultimate laws of physics, no approach has raised higher hopes—or more controversy—than string theory. String theory was first proposed in the 1970s as an attempt to unify all the forces of nature into one coherent framework and, in particular, to bring the force of gravity into the domain of quantum physics. By the early 1990s, however, physicists discovered that string theory suffers from an awkward issue: there are five different string theories. For those advocating that string theory was the unique theory of everything, this was quite an embarrassment. In the mid-1990s researchers started discovering that these different theories—and yet another theory called supergravity—actually describe the same phenomena, giving them some hope that they would amount eventually to a unified theory. The theories are indeed related by what physicists call dualities, which are a kind of mathematical dictionaries for translating concepts back and forth. But, alas, each theory is a good description of phenomena only under a certain range of conditions—for example, at low energies. None can describe every aspect of the universe.

String theorists are now convinced that the five different string theories are just different approximations to a more fundamental theory called M-theory. (No one seems to know what the "M" stands for. It may be "master," "miracle" or "mystery," or all three.) People are still trying to decipher the nature of M-theory, but it seems that the traditional expectation of a single theory of nature may be untenable and that to describe the universe we must employ different theories in different situations. Thus, M-theory is not a theory in the usual sense but a network of theories. It is a bit like a map. To faithfully represent the entire Earth on a flat surface, one has to use a collection of maps, each of which covers a limited region. The maps overlap one another, and where they do, they show the same landscape. Similarly, the different theories in the M-theory family may look very different, but they can all be regarded as versions of the same underlying theory, and they all predict the same phenomena where they overlap, but none works well in all situations.

Whenever we develop a model of the world and find it to be successful, we tend to attribute to the model the quality of reality or absolute truth. But M-theory, like the goldfish example, shows that the same physical situation can be modeled in different ways, each employing different fundamental elements and concepts. It might be that to describe the universe we have to employ different theories in different situations. Each theory may have its own version of reality, but according to model-dependent realism, that diversity is acceptable, and none of the versions can be said to be more real than any other. It is not the physicist's traditional expectation for a theory of nature, nor does it correspond to our everyday idea of reality. But it might be the way of the universe.

About the Authors

Stephen Hawking's work laid the foundation for the modern understanding of black holes and the origin of the universe, although, as he has pointed out, he is at least as famous for his appearances on The Simpsons *and* Star Trek: The Next Generation. *From 1979 to last year, he was Lucasian Professor of Mathematics at the University of Cambridge, the post once held by Isaac Newton. His books include the classic* A Brief History of Time, *which has sold more than nine million copies.*

Leonard Mlodinow is a theoretical physicist at the California Institute of Technology. He has authored seven books–including Euclid's Window: The Story of Geometry from Parallel Lines to Hyperspace *and* The Drunkard's Walk: How Randomness Rules Our Lives*–as well as scripts for* MacGyver *and* Star Trek: The Next Generation.

Record-Breaking Supernova Is Part of a New Class of Objects

By Briley Lewis

The night sky is filled with blips and flashes, a constantly changing sea of lights. Some of these changes are from Earth-bound happenings such as aircraft flying overhead, but some are from distant sources in space. Astronomers hunt for these fleeting phenomena, known as astronomical transients, by observing the sky regularly and looking for differences that appear.

Researchers recently found a transient that outshines all others like it—a supernova known as AT2020mrf. They described their discovery in a paper posted on the preprint server arXiv.org last December and submitted to the *Astrophysical Journal*. They also presented it at a virtual press conference at the 2022 American Astronomical Society meeting in January. This supernova was one of the brightest and most energetic stellar deaths ever seen, and it might provide a rare glimpse into how massive stars give birth to some of the universe's strangest objects: black holes and neutron stars.

The biggest stars die in spectacular fashion. They explode with the energy of a nonillion (that is "1" with 30 zeros behind it) atomic bombs in a supernova, shining so bright that we can sometimes even see them in the night sky with our bare eyes. AT2020mrf was 10,000 times brighter in x-rays than a typical supernova. It followed a few other superbright events that have been observed in recent years—astronomers refer to them as "cow-like" supernovae, or "cows," after the first of their type discovered: AT2018cow. Unlike traditional supernovae, cows shine brightly in high-energy x-rays and radio emission (most supernovae shine brightest in visible light). But AT2020mrf was the brightest cow of the bunch—20 times brighter than the original.

Astronomers Yuhan Yao and Shri Kulkarni, both at the California Institute of Technology, along with Anna Ho of the University of

California, Berkeley, and Daniel Perley of Liverpool John Moores University in England, first spotted this explosion in June 2020 in visible-light images from the Zwicky Transient Facility (ZTF), an automated telescope at California's Palomar Observatory. The bright spot they saw originally seemed like something unremarkable, and the astronomers ignored this event until almost a year later.

In April 2021 Russian scientists collaborating with Yao's group noticed the same event while reviewing data from their Spectrum-Roentgen-Gamma (SRG) x-ray observatory. Their images from July 2020 showed x-rays at the same location as the bright spot in the ZTF data. Upon hearing the news, Yao's mind jumped straight to the mysterious cow-like events because they are the only type of supernova known to emit so many x-rays. Although a year had passed, the initial explosion from AT2020mrf was so incredibly bright that Yao suspected she might still be able to see it with the Chandra X-ray Observatory. Her calculations were correct, and observations showed it clearly, glowing 200 times brighter than the original cow a year after AT2020mrf's explosion.

"It was really very rewarding when SRG saw this source, because we had been thinking that maybe [cow-like events] can be discovered in the x-ray first," Yao says. "This is the first time that one was really discovered in the x-ray."

Astronomers are still puzzling out what makes these supernovae special. The current theory is that cows have a very active "central engine"—something left behind by the star's core, such as a black hole devouring matter or a neutron star rapidly spinning, that provides energy to the supernova. They also have less material shrouding this central region than most exploding stars, providing a view of their interesting center.

The stars that produce cow-like events appear to spew out material as they approach their death, making the environment around them denser and the stars themselves a bit smaller. When they explode, there is less gas in the immediate area around the star's core, allowing x-rays from the central engine to escape. The shockwave the supernova sends through the surrounding space heats up the newly dense environment, too, creating the radio emission astronomers observe. Yao thinks that maybe AT2020mrf was even brighter than most cows because it had thrown off even more mass, allowing the central engine to shine through so brightly in x-rays.

"AT2020mrf is indeed an exciting event, both for what it confirms about the growing class of [cow-like objects] and what it tells us about the diversity of these mysterious stellar explosions," says Brian Metzger, an astrophysicist at Columbia University and the Center for Computational Astrophysics.

With only four previously known cow-like events, AT2020mrf more firmly establishes this group as a new category of explosion. The recent blast also stands out from its classmates, showing the fascinating diversity of stellar deaths. Although scientists understand the broad strokes of how massive stars die, the details are still fuzzy. This is especially true for certain stages of the end of a star's life, such as the silicon burning stage, the last round of fusion a large star can complete, when it fuses silicon together to create iron. This period lasts only around 18 days out of the star's million- to billion-year lifetime. Cows may provide a window into that hard-to-observe time frame and sharpen our understanding of how black holes and neutron stars are born within supernovae.

AT2020mrf and the other cows are also simply thrilling to scientists. "I like the excitement when I just see how one source is different from all others," Yao says. "You know, maybe once in a lifetime, you'll find one of those events, and you need to take action. You need to stay on top of everything to tell the story of the object." Future observations with the Vera C. Rubin Observatory, an upcoming sky-surveying powerhouse, and other telescopes will hopefully give astronomers even more information to work with.

About the Author

Briley Lewis (she/her) is a freelance science writer and Ph.D. Candidate/ NSF Fellow at the University of California, Los Angeles, studying astronomy and astrophysics.

The James Webb Space Telescope Could Solve One of Cosmology's Deepest Mysteries

By Daniel Leonard

On Christmas morning of 2021, astronomers watched their new, greatest tool successfully blast off into space. Now the James Webb Space Telescope (JWST) is fully deployed and has arrived at its deep-space destination, a quiet locale 1.5 million kilometers beyond Earth.

Massimo Stiavelli heads the JWST Mission Office at the institute that allocates research time on the telescope. According to Stiavelli, "every area of science is covered" in the proposals his group has approved, from the search for potentially habitable exoplanets to studies of the earliest stars. Yet he is particularly hopeful that JWST could help settle one of the biggest controversies in modern astronomy: the dispute about the expansion rate of the universe.

"If you try to measure the current expansion rate, well, there's a variety of techniques that people use, and they tend to get a certain number," says Tommaso Treu, an astrophysicist at the University of California, Los Angeles. "And it turns out that those numbers don't match."

Measurements of the universe's expansion rate, known as the Hubble constant, currently cluster around two figures: 67 and 73. Each number is an expression of the same thing–the kilometers-per-second rate of cosmic expansion per every megaparsec (roughly 3.26 million light-years) of space. Although seemingly slight, the difference between these figures is enormous in comparison with the high-precision agreement that exists for other cosmological measurements. Simply put, something is not adding up.

Researchers are not sure how to account for this discrepancy, which they call the Hubble tension. It may just be an error resulting from the different ways the Hubble constant is being measured. Otherwise, the tension could spell trouble for our current understanding of physics, forcing theorists to revisit (and perhaps even discard) some of their most cherished models.

"The quest to measure the expansion rate goes back around 100 years," says Adam Riess, an astrophysicist at Johns Hopkins University. Scientists taking part in that quest fall into two main camps.

The first camp gathers data from the very early universe. These researchers rely on the cosmic microwave background, a residual glow of radiation from roughly 400,000 years after the big bang. By taking measurements from the cosmic microwave background and extrapolating them into the present using our best physical models, astronomers in this camp can reach an estimate for the expansion rate of the universe today. Their calculations indicate that the Hubble constant is around 67.

But Riess adds that JWST will not improve measurements of this kind. Microwaves from the early universe have wavelengths that are too long for JWST–which focuses on infrared light–to detect. Instead JWST has the potential to improve results from the other camp (of which Riess is a prominent member): local measurements.

"Local" is a relative term. Here, it refers to measurements of the Hubble constant that hinge on calculating the distances to stars and galaxies, which may be "only" millions of light-years away. "Measuring distances is what you need to measure the Hubble constant because the Hubble constant is how distances change over time," Treu says.

Astronomers have found a few ways to gauge such celestial distances. Most of them rely on "standard candles," astronomical objects of known brightness. By comparing such an object's actual, intrinsic brightness with its apparent brightness through a telescope, observers can reliably determine its distance from Earth.

Wendy Freedman, an astronomer at the University of Chicago, uses a certain class of red giant stars as her preferred standard candle. "The physics [of these stars] leads to this standard luminosity," which makes them perfect distance indicators, Freedman explains.

Using the Hubble Space Telescope to observe these red giants, Freedman's team arrived at an estimate for the Hubble constant in 2019: roughly 70.

That is on the low end of local estimates. (In fact, it is vexingly midway between the standard values embraced by each camp.) According to Riess, whose standard-candle work uses supernovae and Cepheid variable stars instead of red giants, most local studies have produced somewhat higher values for the Hubble constant–some as high as 75, with an average around 73.

This is a much bigger range than the measurements out of the early-universe camp. Likewise, local studies tend to have greater "error bars" (or uncertainties) than studies that use early-universe data.

That is where JWST can help. By observing in the infrared spectrum, it will be able to look straight through pesky clouds of space dust that all too often interfere with local astronomers' measurements. The Hubble Space Telescope–the previous tool of choice for local astronomers–has far more modest infrared capabilities; its infrared measurements come at the expense of lower image quality. As Riess explains, JWST can do both: observe in infrared and maintain high-resolution imaging.

Crisper, dust-free images: that's the JWST promise.

JWST is such a technological improvement that, rather than altering their methods, many astronomers are planning to carefully replicate their prior research in order to see if the results change. Both Riess and Freedman have been granted research time on JWST to do just that.

Whether or not their results *will* change is uncertain. It is possible that data from JWST could lead local studies to cluster around an estimate for the Hubble constant as low as the one from the early-universe camp. But that seems very unlikely: Riess points out that virtually no local study has produced a result so low, just as no early-universe study has produced a result as high as 73.

So what would it mean if local studies again cluster around 73 but this time with even greater precision? According to Treu, that would imply the Hubble tension is a real discrepancy and not just the result of study error.

If so, Treu adds, that would probably point to something fundamentally absent in our understanding of physics. Because early universe studies rely on physical models to extrapolate their primordial data into the present, missing physics could be the reason those studies are producing a figure as low as 67.

What sort of missing physics? "It could be another neutrino," Riess says. "It could be an early episode of dark energy. It could be decaying dark matter. It could be primordial magnetic fields. All of these have been suggested as things that would help mitigate or explain this." But Riess points out that none of these have "a strong second line of evidence" besides the fact that they could help explain the Hubble tension.

Likewise, Freedman notes that most of these ideas wind up "wrecking" other, agreed on parts of physics somewhere along the line. "This turns out to be really difficult to solve–which is not to say somebody's not going to come up with a brilliant idea at some point," Freedman says.

There may be a hole in physics. There is no guarantee that JWST will help us figure out how to fill it. But by giving greater insight into the Hubble tension, JWST can at least help confirm that the hole is really there.

About the Author

Daniel Leonard is a freelance science journalist and former Scientific American *editorial intern whose work focuses on space, tech and natural history.*

10 Years after the Higgs, Physicists Are Optimistic for More Discoveries

By Marcela Carena

Imagine that you have just arrived on a planet in another solar system. Suddenly, five minutes after you landed, you spot an alien life-form. This is an amazing discovery! You may well spend decades trying to understand this exotic being, probing its properties and investigating how it came to be there. At the same time, you expect that there may be other fascinating creatures around, maybe even more intriguing than the first and possibly much harder to get a glimpse of.

This is how it feels for particle physicists as we begin a new phase, called Run 3, at the world's most powerful particle accelerator: the Large Hadron Collider (LHC) at CERN near Geneva. This month marks the 10th anniversary of the discovery of the Higgs boson, a long-sought particle that had been predicted almost 50 years earlier. The LHC was built to find the Higgs boson and it did. Its next goal is to find clues to help us decipher other unresolved mysteries. Although the machine has not yet uncovered other novel fundamental particles—especially the hoped for supersymmetric particles that popular theories predicted and may still be out there—since the Higgs boson, the future at the LHC is promising. We have many new avenues to explore and many reasons for optimism.

The Higgs boson discovery, which came just four years after the LHC opened, was a lucky strike—it could have taken much longer to detect the particle, or we might never have found it at all. The particle's mass might not have been in the range accessible at the collider, or it might not have interacted enough with other particles to be produced in the LHC collisions. It might not even have existed at all. And nature was even kinder: for reasons that we don't yet understand, it arranged for the Higgs boson's mass to be 125 times the mass of a proton, a value that causes the Higgs to decay into many of the particles we know at similar rates.* This property makes it convenient to explore how the Higgs boson talks to these other particles and opens many opportunities to look for the unexpected.

The Higgs boson was the remaining piece of the Standard Model of particle physics, our leading theory of the properties and interactions among the fundamental bits of nature. Much of particle physics does not fit into this model, however. The current state of our field feels like trying to understand the science of cooking when all you have is a good grasp on the theory of how water boils. The Standard Model is silent about dark matter and even the force of gravity. Neutrinos are there, but their tiny masses are unaccounted for. Ordinary matter is there but with no explanation of how it prevailed over antimatter after the big bang. The Higgs boson is there but with no attempt to explain why the invisible Higgs energy field turned itself on in the early universe to give mass to other particles—or why their masses are as different as those of an ant and a whale or why the Higgs gave itself a mass that puts the present-day universe at the edge of cosmic instability.

The LHC was designed as a discovery machine to help us answer these questions, and fortunately for us, it has another 20 years on its horizon before it shuts down. The flagship detectors at the collider—ATLAS and CMS—have become quite different experiments in Run 3 than they were 10 years ago. Both have received upgraded technologies, and a new generation of talented scientists is pursuing novel ideas for how to glimpse what may be out there. In Run 3, all of the LHC experiments will be pushing into previously unexplored territory on multiple fronts. I am trembling with excitement about what may lie ahead.

We have already been able to produce many thousands of Higgs bosons at the LHC and are now working toward detecting the rarer ways by which the particle can be produced and then decay into other particles. There are plenty of chances for surprises, either in precision measurements that may show the Higgs is produced or decays somewhat differently from our Standard Model predictions or through the observation of exotic phenomena related to the Higgs. For instance, the Higgs boson might decay into dark matter, or Higgs decays may violate the expected symmetry between matter and antimatter.

So far we have only seen collisions that produce a single Higgs boson at a time. But we think that it should also be possible to produce two Higgs bosons in one collision. This "di-Higgs" production would give us a direct window into how the Higgs energy field turned on after the big bang because it is a direct measure of how strongly the Higgs boson, and therefore the Higgs energy field, interacts with itself. The Standard Model predicts that collisions producing two Higgs bosons will occur at a finite but tiny rate, suggesting that this process will become detectable near the end of the LHC's lifetime. This is an exciting prospect, but there is also no compelling reason to believe the details of this prediction: the Standard Model does not claim to know the origins of the Higgs boson or understand the mechanisms of the invisible Higgs field in the early universe. A di-Higgs signal could potentially be seen sooner, during Run 3, perhaps induced by new particles that enhance the process.

Less than one LHC collision in a billion produces a Higgs boson, so the initial discovery was like finding a needle in a very large haystack. Today theorists have proposed many possibilities for other alien particles that could eventually appear in our detectors. But the challenge now is akin to searching for *something* in a haystack when you don't even know if you are seeking a needle or some other object entirely.

Other new opportunities abound. Ten years ago most physicists would have dismissed the idea, if anyone had been crazy enough to suggest it, that we could attach a neural network to an LHC detector to analyze its findings. Today, thanks to the innovations developed by a brilliant cadre of my junior colleagues and industry partners eager to push the boundaries of artificial intelligence, a network looks at 40 million LHC collisions per second to decide what appears interesting enough to record for later study by humans.When you see photographs of the colossal ATLAS and CMS detectors, most of the detectors' volume that you are looking is their outer portion, which was designed to detect and measure one special kind of particle called the muon–a cousin to the electron that is often produced when particles decay, including the Higgs boson. Muons penetrate through matter more easily than other particles that are stopped and measured in the inner portions of the detectors. Within just the past two years, innovators at both CMS and ATLAS have realized that they can repurpose the outer detectors for potential discoveries that would otherwise be missed. Many models that try to explain dark matter propose the existence of exotic long-lived particles that may penetrate through the inner detectors and arrive in the outer detectors before decaying to standard particles. This scenario would produce a discovery signal that could be relatively easy to see–but only if you knew to look for it.

Ten years after the Higgs boson's discovery, the field of particle physics is blooming with new ideas aimed at shedding light on profound mysteries. The Large Hadron Collider is starting a new chapter in its life, with more powerful particle beams, enhanced detector capabilities and more sophisticated techniques to enable discovery. Both experimentalists and theorists are expanding the frontiers of their ingenuity in this thrilling journey of exploration.

This is an opinion and analysis article, and the views expressed by the author or authors are not necessarily those of Scientific American.

**Editor's Note (7/6/22): This sentence was edited after posting to correct the description of the Higgs decaying into many particles at similar rates.*

About the Author

Marcela Carena is a particle physicist and head of the Theory Division at Fermi National Accelerator Laboratory in Batavia, Ill., and a professor of physics at the University of Chicago, where she is a member of the Enrico Fermi Institute and the Kavli Institute for Cosmological Physics.

GLOSSARY

anomalous Inconsistent with or deviating from what is normal or expected.

dark matter Matter not yet detected by astronomers that is thought to exist to account for various observed gravitational effects.

entropy A measurement of the energy that is present in a system or process but is not available to do work.

event horizon The boundary of a black hole beyond which nothing can escape.

multiverse A theoretical reality that includes a possibly endless number of parallel universes.

neutrino A subatomic particle that has no charge and is believed to have a very small mass.

paradox A statement containing two opposite ideas that make it seem impossible or unlikely.

quantum theory A theory in physics based on the idea that energy is made of small separate units of energy.

redshift A displacement of the light of a celestial body toward longer wavelengths that is caused by the presence of a gravitational field.

singularity A region of infinite mass density at which space and time are infinitely distorted by gravitational forces.

string theory A theory in physics that states that all elementary particles are manifestations of the vibrations of one-dimensional strings.

superposition The combination of two distinct physical phenomena of the same type so that they exist as part of the same event.

FURTHER INFORMATION

Feldman, Andrey. "String theory may offer an alternative to dark energy," Advanced Science News, March 6, 2023, https://www.advancedsciencenews.com/string-theory-may-offer-an-alternative-to-dark-energy/.

Greenfieldboyce, Nell. "Scientists have found signs of a new kind of gravitational wave. It's really big," National Public Radio, June 28, 2023, https://www.npr.org/2023/06/28/1183878163/new-gravitational-wave-supermassive-black-hole.

Lea, Robert. "Supermassive black holes grow surprisingly quickly, study suggests," Space.com, June 28, 2023, https://www.space.com/supermassive-black-holes-grow-surprisingly-quickly.

Lea, Robert. "The Upcoming Superhero Movie The Marvels Plays On the Theory of Quantum Entanglement. Here's How It Works," *Popular Mechanics*, June 28, 2023, https://www.popularmechanics.com/science/a44373417/the-marvels-movie-how-quantum-entanglement-works/.

Padavic-Callaghan, Karmela. "Exotic cosmic objects in string theory may look like leaky black holes," NewScientist, April 28, 2023, https://www.newscientist.com/article/2370625-exotic-cosmic-objects-in-string-theory-may-look-like-leaky-black-holes/.

Schultz, Isaac. "Decoding Dark Matter and Energy: What You Need to Know About the Euclid Mission Launching This Weekend," Gizmodo, June 28, 2023, https://gizmodo.com/euclid-telescope-mission-explainer-dark-matter-energy-1850585534.

Starr, Michelle. "Physicists Conduct The Most Massive Test Ever of The Eintein-Podolsky-Rosen Paradox," ScienceAlert, June 12, 2023, https://www.sciencealert.com/physicists-conduct-the-most-massive-test-ever-of-the-einstein-podolsky-rosen-paradox.

Weitering, Hanneke. "How Theoretical Physicists Can Help Find Aliens," *Scientific American*, June 28, 2018, https://www.scientificamerican.com/article/how-theoretical-physicists-can-help-find-aliens/.

CITATIONS

1.1 The Universe Began with a Bang, Not a Bounce, New Studies Find by James Riordon (May 24, 2023); 1.2 Colliding Dwarf Galaxies Reveal a Glimpse of the Early Universe by Meghan Bartels (March 3, 2023); 1.3 Ancient Galaxy Clusters Offer Clues about the Early Universe by Arianna S. Long (January 1, 2021); 1.4 Giant Galaxies from the Universe's Childhood Challenge Cosmic Origin Stories by Robin George Andrews (January 27, 2021); 1.5 JWST's First Glimpses of Early Galaxies Could Break Cosmology by Jonathan O'Callaghan (September 14, 2022); 2.1 Black Hole Discovery Helps to Explain Quantum Nature of the Cosmos by Edgar Shaghoulian (September 1, 2022); 2.2 Colliding Supermassive Black Holes Discovered in Nearby Galaxy by Allison Parshall (January 18, 2023); 2.3 Did Astronomers Just Discover Black Holes from the Big Bang? by Nola Taylor Tillman (January 13, 2020); 2.4 Astronomers Spy a Giant Runaway Black Hole's Starry Wake by Kenna Hughes-Castleberry (April 10, 2023); 2.5 Black Holes, Quantum Entanglement and the No-Go Theorem by Zoë Holmes, Andrew Sornborger (July 4, 2021); 2.6 Neutrinos from a Nearby Galaxy Reveal Black Hole Secrets by Phil Plait (December 20, 2022); 3.1 Could Gravity's Quantum Origins Explain Dark Energy? by Conor Purcell (October 28, 2021); 3.2 Astronomers Might See Dark Matter by Staring into the Void by Anil Ananthaswamy (June 8, 2022); 3.3 Dark Matter Hunters Need Fresh Answers by Joseph Howlett (April 5, 2023); 3.4 Stars Made of Antimatter Might Be Lurking in the Universe by Leto Sapunar (June 7, 2021); 3.5 Are Telescopes the Only Way to Find Dark Matter? by Chanda Prescod-Weinstein (April 1, 2022); 4.1 String Theory May Create Far Fewer Universes Than Thought by Clara Moskowitz, Lee Billings (July 30, 2018); 4.2 Tangled Up in Spacetime by Clara Moskowitz (October 26, 2016); 4.3 Found: A Quadrillion Ways for String Theory to Make Our Universe by Anil Ananthaswamy (March 28, 2019); 4.4 Quantum Entanglement Isn't All That Spooky After All by Chris Ferrie (February 13, 2023); 5.1 Tiny Gravitational-Wave Detector Could Search Anywhere in the Sky by Tim Folger (June 30, 2020); 5.2 Gravitational Waves Probe Exotic Matter inside Neutron Stars by Clara Moskowitz (December 17, 2020); 5.3 Galaxy-Size Gravitational-Wave Detector Hints at Exotic Physics by Adam Mann (February 3, 2021); 6.1 The Elusive Theory of Everything by Stephen Hawking, Leonard Mlodinow (October 1, 2010); 6.2 Record-Breaking Supernova Is Part of a New Class of Objects by Briley Lewis (February 21, 2022); 6.3 The James Webb Space Telescope Could Solve One of Cosmology's Deepest Mysteries by Daniel Leonard (January 25, 2022); 6.4 10 Years after the Higgs, Physicists Are Optimistic for More Discoveries by Marcela Carena (July 5, 2022

Each author biography was accurate at the time the article was originally published.

INDEX